INTRODUCTION TO THE THEORY OF NONLINEAR ELLIPTIC EQUATIONS

Introduction to the Theory of Nonlinear Elliptic Equations

by

Jindřich Nečas
University of Prague, Czechoslovakia

A Wiley – Interscience Publication

JOHN WILEY & SONS
Chichester · New York · Brisbane · Toronto · Singapore

British Library Cataloguing in Publication Data:

Nečas, Jindřich
Introduction to the theory of nonlinear elliptic equations.
1. Differential equations, Elliptic
2. Differential equations, Nonlinear
I. Title
515.3′53 QA377
ISBN 0 471 90894 0

Library of Congress Cataloging-in-Publication Data:

Nečas, Jindřich.
Introduction to the theory of nonlinear elliptic equations.
Bibliography: p.
Includes index.
1. Differential equations, Elliptic. I. Title.
QA377.N39 1986 515.3′53 85–22693
ISBN 0 471 90894 0

Printed in the German Democratic Republic.

Preface

These lecture notes are a very free continuation of my book "Les méthodes directes en théorie des équations elliptiques", Prague–Paris, 1967, in the nonlinear case. The realization of the existence-regularity scheme for nonlinear systems had to be preceded by the effort of mathematicians from the whole world. Finally, in last years, some new ideas have appeared that enabled us to present a theory complete in some sense. I underline that I omit completely the spectral, bifurcation, multiplicity, genericity, and other problems that belong rather to functional analysis then to the theory of elliptic differential equations. Nevertheless, in Chapters 3 and 4, which are concerned with the existence of solution and with approximate methods, many subjects have the same character. The main part of the lecture notes is Chapters 5 and 6 on the regularity questions. I added some applications to elasticity, which are far from being immediate and which show in a large extent some fundamental questions remaining still open.

A lot of the topics of these lecture notes were discussed in the seminar on partial differential equations in the Mathematical Institute of the Czechoslovak Academy of Sciences and in the lectures that I held at the Faculty of Mathematics and Physics of the Charles University, at Scuola Normale Superiore di Pisa and at the University of Pierre et Marie Curie in Paris.

It is the author's pleasant duty to thank all his colleagues and friends for variable discussions: O. A. Oleĭnik, S. Campanato, M. Giaquinta, P. Ciarlet, G. Tronel, J. Frehse, A. Kufner, J. Stará, O. John, R. Švarc, M. Krbec, M. Šilhavý and P. Drábek. I also wish to thank K. Segeth for his help with the English translation, R. Pachtová for her excellent typing, and the TEUBNER Publishing House, Leipzig, for their collaboration and patience.

Prague, May 1982 J. Nečas

Contents

Chapter 1

The topic of the lecture notes and something on modelling by partial differential equations

1.1. Introduction

The significance of nonlinear elliptic partial differential equations grows up both from the theoretical point of view and as a consequence of their numerous applications. Taking into account the main results, the classical as well as the most recent ones, we can affirm that the theory of these equations is just creating a harmonic entirety where the basic questions are answered. These lecture notes have been written as an introduction though they are in some directions complete enough. I will not use the introduction to describe the history of the development of nonlinear partial differential equations and I recommend the reader the books by O. A. Ladyženskaja, N. N. Uralceva [1], Ch. B. Morrey [2], J. L. Lions [3], D. Gilbarg, N. S. Trudinger [4], S. Fučík, A. Kufner [5] and E. Giusti [6]. We shall touch some historical features in the sequel.

In these lecture notes, the study of second order systems in the divergence form

$$-\frac{\partial}{\partial x_i}[a_i^r(x, u, \nabla u)] + a^r(x, u, \nabla u) = -\frac{\partial f_i^r}{\partial x_i} + f^r \tag{1.1.1}$$

will be in the centre of our interest. Here the summation over the repeated subscript is understood and

$$x = (x_1, x_2, \ldots, x_n) \in \mathbb{R}^n, \qquad u = (u_1, u_2, \ldots, u_m),$$

$$\nabla u = (\nabla u_1, \nabla u_2, \ldots, \nabla u_m), \qquad \nabla u_r = \left(\frac{\partial u_r}{\partial x_1}, \frac{\partial u_r}{\partial x_2}, \ldots, \frac{\partial u_r}{\partial x_n}\right).$$

One comes to such systems, for example, in the study of the critical points of the functionals

$$\int_{\Omega} F(x, u, \nabla u)\, dx . \tag{1.1.2}$$

Elliptic partial differential equations, together with boundary conditions, are models for many physical, mechanical, and technical phenomena and we shall touch such models in these lecture notes in order to clarify that the construction of modern models as well as the study of these modern models is considerably neglected as compared with the study of the classical ones where very subtle results have often nearly nothing in common with the reality. Such a trivial case is, for example, the functional of the total potential energy of a membrane

$$\Phi(u) \stackrel{\text{def}}{=} \int_{\Omega} \left[T(x) \left(\sqrt{1 + \sum_{i=1}^{2} \left(\frac{\partial u}{\partial x_i} \right)^2} - 1 \right) - uf \right] dx , \tag{1.1.3}$$

which is correct only in the case of $T(x) = \text{const}$ because otherwise one does not obtain the conditions of equilibrium for the momentum, as we shall see in 1.2. In this case one must study a parametrical "minimal surface" problem. Also the usual trick

$$\sqrt{1 + \sum_{i=1}^{2} \left(\frac{\partial u}{\partial x_i} \right)^2} - 1 \doteq \frac{1}{2} \sum_{i=1}^{2} \left(\frac{\partial u}{\partial x_i} \right)^2 \tag{1.1.4}$$

leads to the Poisson equation $-\dfrac{\partial}{\partial x_i}\left(T \dfrac{\partial u}{\partial x_i} \right) = f$ the theory of which completely differs from the theory of the equation for the surface with prescribed mean curvature $-\dfrac{1}{T} f$ (if T is constant).

The main questions concerning the systems (1.1.1), if we add some boundary conditions (for example if we prescribe $u = h$ on the boundary $\partial\Omega$ of the considered domain $\Omega \subset \mathbb{R}^n$), are

(i) the existence and uniqueness of the solution; if the uniqueness does not occur, then the structure of the solutions, generic properties, and the bifurcation,

(ii) approximate methods for finding the solution,

(iii) the regularity of solution,

(iv) what such systems can model.

In these lecture notes we shall not be concerned with systems of higher order; roughly speaking, those differ from (1.1.1) by more indices. We shall make only brief remarks and some references to this topic. In these lecture notes, we shall be interested both in variational methods and in methods of monotone operators. Nowadays, these are the classical results and the reader can consult also other lecture notes of this series, see S. Fučík, J. Nečas, J. Souček [7] and E. Zeidler [8].

The main point of these lecture notes is the solvability of elliptic systems and the qualitative behaviour of their solutions. Hence all the functional-analytic methods are presented from this point of view.

Surveying the historical development of the theory of nonlinear elliptic equations, we find that one of the fundamental steps was the introduction of the Sobolev spaces $W^{k,p}(\Omega)$, see S. L. SOBOLEV [9], the development of the abstract variational methods, see for example M. M. VAJNBERG [10], and the methods of monotone operators, see J. L. LIONS [3]. Let us mention that the fundamentals of those direct methods were given much earlier; if we will not go back to works of RIEMANN and DIRICHLET we can find them in the book by R. COURANT and D. HILBERT [11]. The works of F. RELLICH, E. TREFFTZ, S. L. SOBOLEV, S. G. MICHLIN, R. CACCIOPPOLI, C. MIRANDA, L. SCHWARTZ, L. GÅRDING, K. O. FRIEDRICHS, and others have established the fundamentals of the theory of weak solutions to elliptic equations and this theory was accomplished, especially for nonlinear equations, by M. M. VAJNBERG, G. MINTY, J. LERAY, M. I. VIŠIK, J. L. LIONS, F. E. BROWDER, H. BRÉZIS and others. The Sobolev spaces $[W^{1,p}(\Omega)]^m$ are the spaces of vector functions $u = (u_1, u_2, \ldots, u_m)$ that are L^p-integrable in the considered domain along with their first derivatives.

The solvability of the systems (1.1.1) in $[W^{1,p}(\Omega)]^m$ requires some growth conditions for coefficients, for example

$$(1.1.5) \qquad |a_i^r(x, u, \nabla u)| + |a^r(x, u, \nabla u)| \leqq c(1 + |u| + |\nabla u|)^{p-1}.$$

On the other hand, when such systems are models (physical, technical, ...), the response functions a_i^r, a^r are defined for u and ∇u only in some bounded regions, and the conditions (1.1.5) are fictitious extrapolations. This incongruity can be eliminated if one considers solutions from $[W^{1,\infty}(\Omega)]^m$. In general, it is not known whether such solutions do exist under reasonable conditions. One can go further and look for solutions in $[C^1(\bar{\Omega})]^m$ or $[C^{1,\alpha}(\bar{\Omega})]^m$ and, as we shall see later, this is just the question (iii), which is also the focal point of these lecture notes. There are books or lecture notes on the questions (i), (ii), and (iv) that are more complete than these lecture notes and the author added this topic rather because of the integrity of the material presented.

As far as the point (ii) is concerned, it does not surpass a standard treatment too much. For more details, the reader is recommended to consult the book by J. CÉA [12]. We look for the steepest descent methods and, also from this point of view, for the Newton method. In general, this method requires the C^1 regularity of solutions, which is not generally known. Surprisingly, the imbedding method requires practically only weak solutions. (The imbedding method is, roughly speaking, a continuous analogue of the Newton method.)

The central point (iii) of these lecture notes is the $[C^{1,\alpha}(\Omega)]^m$ or $[C^{1,\alpha}(\bar{\Omega})]^m$ regularity of weak solutions. In general under standard assumptions on the systems (1.1.1) and for a weak solution from $[W^{1,p}(\Omega)]^m$ (a solution in the sense of distributions), there exists a set M, closed in Ω, of zero measure (more precisely, see later), and such that the solution is from $[C^{1,\alpha}(\Omega \setminus M)]^m$. This is the so-called partial regularity. Under some further assumptions that are sufficient and "necessary", in principle the same method "of partial regularity" gives that $M = \emptyset$. This condition is for the interior regularity, i.e. for proving that a Lipschitz continuous solution to (1.1.1) lies in $[C^{1,\alpha}(\Omega)]^m$ (which means that it is Hölder continuous on every compact $K \subset \Omega$) denoted as $L(\mathbb{R}^n)$—the Liouville-type condition.

For the system (1.1.1) it means:

$\forall x^0 \in \Omega$, $\forall \xi \in \mathbb{R}^m$ the solutions v to the system

$$-\frac{\partial}{\partial x_i}[a_i^r(x^0, \xi, \nabla v)] = 0 \tag{1.1.6}$$

in R^n with a bounded gradient $|\nabla v(x)| \leqq c < \infty$ are polynomials of at most first degree. It is known that $C^{1,\alpha}(\Omega)$ regularity holds for the dimension $n = 2$ and for $m = 1$, $n \geqq 2$, and we reprove these results in Chapters 5 and 6. The regularity of weak solutions to (1.1.1) was proved for $n = 2$ by CH. B. MORREY [13], for $m = 1$, $n \geqq 2$ by E. DE GIORGI [14], and J. NASH [14¹] and the condition (1.1.6) was discovered by M. GIAQUINTA, J. NEČAS [15]. There are many important steps in the regularity problem and we removed all the details to Chapters 5 and 6. Let us mention that we shall suppose the system (1.1.1) to be *very strongly* elliptic throughout these lecture notes, i.e. such that

$$\frac{\partial a_i^r}{\partial \eta_j^s}(x, \xi, \eta)\, \zeta_i^r \zeta_j^s > 0 \quad \text{for} \quad \zeta \neq 0. \tag{1.1.7}$$

Also, under this condition, there exists system (1.1.1) with real-analytic coefficients for which the condition (1.1.6) is not satisfied, namely systems with the solution $u^{rs} = \dfrac{x_r x_s}{|x|} - \dfrac{1}{n}\delta_{rs}|x|$, $r, s = 1, 2, \ldots, n$, for $n \geqq 3$, see M. GIAQUINTA, J. NEČAS [16].

Nevertheless, we shall close this introduction by an important historical remark. The regularity problem in words of the analyticity of the extremum of the functional (1.1.2) with $F(x, \xi, \eta)$ analytic was formulated by D. HILBERT in 1900 as his 19th problem and the way of its solution connected with the names of S. N. BERNSTEIN, J. LERAY, J. SCHAUDER, H. WEYL and other mathematicians may serve as a beautiful description of the development of mathematics.

1.2. Partial differential equations in modelling

In modelling, one supposes very often that response functions and quantities to be considered are differentiable enough. On the other hand, the mathematical formulation is often given in terms of generalized solutions. The connection between these two points of view is not, up to this time, clarified enough.

Less often the formulation corresponds to the mathematical model. It will be very useful to consider also this way in modelling.

1.2.1. EXAMPLE. Everybody knows the heat conduction equation. Let us derive it once again.

Let Ω be the considered body. We suppose that Ω is a domain in $\mathbb{R}^3$. But it is more reasonable to consider $\bar{\Omega}$. Is it true that $\partial\Omega$ is always insignificant? Let us suppose the time running through the interval $[0, \infty)$. Let $u(t, x)$ be the temperature of Ω at the point (t, x); let us accept this point of view. Let $D \subset \Omega$ be a subdomain and let $S \subset \partial D$. We suppose that the Newton law for the heat flux is valid: in the time interval $[t_1, t_2]$ the heat flux through S from $\Omega \setminus D$ to D is

$$q_{[t_1, t_2]} = + \int_{t_1}^{t_2} \mathrm{d}t \int_S k(x, u, \nabla u) \frac{\partial u}{\partial n} \, \mathrm{d}S, \tag{1.2.2}$$

where $k(x, u, \nabla u)$ is the heat conductivity. One writes (1.2.2) also in the form

$$\mathrm{d}q = -k \frac{\partial u}{\partial n} \mathrm{d}t \, \mathrm{d}S. \tag{1.2.3}$$

How many assumptions are hidden in (1.2.3)! Clearly first (1.2.3) must make sense; then $\mathrm{d}q$ is a sign measure on $[t_1, t_2] \times \partial D$. Of course, a surface measure on ∂D must be defined. Further (1.2.3) implies that the measure $\mathrm{d}q$ has density $-k \dfrac{\partial u}{\partial n}$ that depends linearly on $\dfrac{\partial u}{\partial n}$. So ∂D must be an oriented surface. On the other hand, $k = k(x, u(t, x), \nabla u(t, x))$: the principle of locality ...

The reader sees from these simplest remarks that (1.2.3) is not obvious and, under some conditions, it can be false. One way continue with the classical balance of heat flow. The increase of the temperature of the body D during the time $[t_1, t_2]$ needs the heat

$$\int_{t_1}^{t_2} \int_D \varrho(x) \, c(x, u) \frac{\partial u}{\partial t} (t, x) \, \mathrm{d}t \, \mathrm{d}x, \tag{1.2.4}$$

where ϱ is the density and c is the specific heat. (One can make as many remarks as before.) If there is a heat source in Ω, it yields the heat

$$\int_{t_1}^{t_2} \int_D f(t, x) \, \mathrm{d}t \, \mathrm{d}x \tag{1.2.5}$$

during the time $[t_1, t_2]$ in D.

The total heat flux from $\Omega \setminus D$ to D is

$$\int_{t_1}^{t_2} \int_{\partial D} k \frac{\partial u}{\partial n} \, \mathrm{d}t \, \mathrm{d}S. \tag{1.2.6}$$

Hence,

$$\int_{t_1}^{t_2} \int_{\partial D} k \frac{\partial u}{\partial n} \, \mathrm{d}t \, \mathrm{d}S + \int_{t_1}^{t_2} \int_D f \, \mathrm{d}t \, \mathrm{d}x = \int_{t_1}^{t_2} \int_D \varrho c \frac{\partial u}{\partial t} \, \mathrm{d}t \, \mathrm{d}x. \tag{1.2.7}$$

From Green's formula we have

$$\int_{t_1}^{t_2}\int_D \left(\frac{\partial}{\partial x_i}\left[k\frac{\partial u}{\partial x_i}\right] + f\right) \mathrm{d}t\,\mathrm{d}x = \int_{t_1}^{t_2}\int_D \varrho c\frac{\partial u}{\partial t}\,\mathrm{d}t\,\mathrm{d}x; \tag{1.2.8}$$

we let the reader formulate the assumptions for the validity of (1.2.8). If D is regularly shrinking to a point x and $t_1 = t$, $t_2 \to t$, then, if the integrand is continuous, for example, the classical heat-conduction equation follows from (1.2.8).

If the temperature approaches a steadyvalue for $t \to \infty$, i.e. if $u(t, x) \to U(x)$ in some sense, it may be excepted that

$$-\frac{\partial}{\partial x_i}\left[k(x, U, \nabla U)\frac{\partial U}{\partial x_i}\right] = F(x) \quad \text{in } \Omega \tag{1.2.9}$$

and, for example,

$$U(x) = H(x) \quad \text{on } \partial\Omega, \tag{1.2.10}$$

where H is prescribed and $F(x) = \lim_{t\to\infty} f(t, x)$. The classical formula for the Cauchy problem

$$\frac{\partial u}{\partial t} = \Delta u \quad \text{in} \quad \mathbb{R}_+^{n+1} = \{(t, x) \mid t > 0\}, \tag{1.2.11}$$

$$u(0, x) = \phi(x): \tag{1.2.12}$$

$$u(t, x) = \frac{1}{2^n\pi^{n/2}t^{n/2}}\int_{R^n} \exp\left(-\frac{|x-y|^2}{4t}\right)\phi(y)\,\mathrm{d}y \tag{1.2.13}$$

gives that the heat flow has an infinite speed, which is impossible. Where is the mistake?

1.2.14. EXAMPLE. Let us consider a membrane M and let us suppose first that it is a surface given in $\mathbb{R}^3$, as usually, by

$$y_1 = x_1, \qquad y_2 = x_2, \qquad y_3 = u(x_1, x_2), \tag{1.2.15}$$

where $(x_1, x_2) \in \Omega \subset \mathbb{R}^2$. Let us consider first the steady state; let us suppose that a surface force $f_3(x_1, x_2)$ (in the direction of the axis x_3) related to the unit volume of Ω is acting on the membrane. The total work of the surface forces, if we start from the configuration $u_0(x_1, x_2)$, therefore is

$$\int_\Omega f_3(u - u_0)\,dx_1\,dx_2. \tag{1.2.16}$$

Let us consider a part S of the membrane M the projection of which is $O \subset \Omega$ with ∂O smooth enough. Let us suppose that we can define a tension vector σ at the points $(x_1, x_2) \in \partial O$, related to the unit surface of ∂O, such that $\sigma = \sigma((x_1, x_2), \nu)$, where ν is the outer normal to ∂O at the point (x_1, x_2). We suppose that the vector σ lies in the plane tangent to the surface. If we put $p = \sqrt{1 + \left(\frac{\partial u}{\partial x_1}\right)^2 + \left(\frac{\partial u}{\partial x_2}\right)^2}$ then the unit normal to the surface is

(1.2.17) $$n = \left(-\frac{u_{x_1}}{p}, -\frac{u_{x_2}}{p}, \frac{1}{p}\right).$$

Hence

(1.2.18) $$(n, \sigma) = 0.$$

If t is the tangent vector to ∂S at the point y, then we suppose

(1.2.19) $$(t, \sigma) = 0.$$

So we can suppose that the tension vector $g(x_1, x_2)$ acts on some part of "∂S" the projection of which is $\Gamma \subset \partial\Omega$. The total work of this tension vector is

(1.2.20) $$\int_\Gamma g_3(u - u_0)\, \mathrm{d}s.$$

Let us suppose that the total potential energy of tension vectors is

(1.2.21) $$\int_\Omega T(x_1, x_2)\, [p - p_0]\, \mathrm{d}x_1\, \mathrm{d}x_2,$$

which means that the increment $\mathrm{d}W$ of the stored energy of the membrane is proportional to the change of the surface of the membrane. So the functional of the total potential energy of the membrane (apart from a constant) is

(1.2.22) $$\Phi(u) = \int_\Omega (Tp - f_3 u)\, \mathrm{d}x_1\, \mathrm{d}x_2 - \int_\Gamma g_3 u\, \mathrm{d}s.$$

Let us consider (formally) the minimum of $\Phi(u)$ over the set of u such that $u = u^0$ on $\partial\Omega \setminus \Gamma$. Let u be such a minimum. If $h = 0$ on $\partial\Omega \setminus \Gamma$, it follows that

(1.2.23) $$\begin{aligned} D\Phi(u, h) &= \frac{\mathrm{d}}{\mathrm{d}t}[\Phi(u + th)]_{t=0} \\ &= \int_\Omega \left[\frac{T}{p}(\nabla u \cdot \nabla h) - f_3 h\right] \mathrm{d}x_1\, \mathrm{d}x_2 \\ &\quad - \int_\Omega g_3 h\, \mathrm{d}s = 0. \end{aligned}$$

If $O \subset \bar{O} \subset \Omega$, then it follows in the same manner that

(1.2.24) $$\int_O \left[\frac{T}{p}\nabla u\, \nabla h - f_3 h\right] \mathrm{d}x_1\, \mathrm{d}x_2 - \int_{\partial O} \sigma_3 h\, \mathrm{d}s = 0.$$

So we get *Euler's equation*

(1.2.25) $$\frac{\partial}{\partial x_i}\left(\frac{T}{p}\frac{\partial u}{\partial x_i}\right) + f_3 = 0 \quad \text{in } \Omega$$

and

(1.2.26) $$\frac{T}{p}\frac{\partial u}{\partial n} = g_3 \quad \text{on } \Gamma,$$

where $\frac{\partial u}{\partial n}$ is the derivative with respect to the outer normal to $\partial\Omega$. It follows from (1.2.18), (1.2.19) and (1.2.24) that

(1.2.27) $$\sigma_3 = \frac{T}{p}\frac{\partial u}{\partial n}$$

holds on ∂O, hence the conditions (1.2.18), (1.2.19) give finally

$$\sigma_1 = \left(\frac{\partial x_2}{\partial s} + \frac{\partial u}{\partial x_2}\frac{\partial u}{\partial s}\right)\frac{T}{p}, \qquad \sigma_2 = -\left(\frac{\partial x_1}{\partial s} + \frac{\partial u}{\partial x_1}\frac{\partial u}{\partial s}\right)\frac{T}{p}.$$
(1.2.28)

It follows from (1.2.24) that the resultant of forces on O is zero only in the x_3 direction, because $\int_{\partial O} \sigma_1 \, ds \neq 0 \neq \int_{\partial O} \sigma_2 \, ds$ in general. If we put $p = 1$ in (1.2.25)–(1.2.28), we get

(1.2.29) $$\frac{\partial}{\partial x_i}\left(T\frac{\partial u}{\partial x_i}\right) + f_3 = 0 \quad \text{in } \Omega,$$

(1.2.30) $$T\frac{\partial u}{\partial n} = g_3 \quad \text{on } \Gamma,$$

(1.2.31) $$u = u_0 \quad \text{on } \partial\Omega \setminus \Gamma,$$

(1.2.32) $$T\frac{\partial u}{\partial n} = \sigma_3, \qquad T\frac{dx_2}{ds} = \sigma_1, \qquad -T\frac{dx_1}{ds} = \sigma_2.$$

Also in this linearized case, the condition of equilibrium for forces is satisfied in the x_3 direction and on the direction of the axes x_1 and x_2 only if $T = \text{const}$. In the case of $T = \text{const}$ we obtain also the equilibrium of momentum on O (we put $f = (0, 0, f_3)$):

(1.2.33) $$\int_O (y \times f) \, dO + \int_{\partial O} (y \times \sigma) \, ds = 0,$$

where $\times$ denotes the vector product. We leave the verification to the reader.

Hence, what is the exact model of a membrane, provided that it is a smooth two-dimensional surface with the reference domain Ω? It is natural to consider a general displacement vector $u = (u_1, u_2, u_3)$ and to suppose

(1.2.34) $$y_1 = x_1 + u_1(x_1, x_2), \qquad y_2 = x_2 + u_2(x_1, x_2),$$
$$y_3 = u_3(x_1, x_2).$$

The initial position is given by the vector function $y^0(x_1, x_2)$. If

$$E = \left(\frac{\partial y}{\partial x_1}, \frac{\partial y}{\partial x_1}\right), \qquad G = \left(\frac{\partial y}{\partial x_2}, \frac{\partial y}{\partial x_2}\right), \qquad F = \left(\frac{\partial y}{\partial x_1}, \frac{\partial y}{\partial x_2}\right),$$
(1.2.35)

then

(1.2.36) $$dW = T\left[\sqrt{EG - F^2} - \sqrt{E_0 G_0 - F_0^2}\right] d\Omega$$

and the functional of the total potential energy (apart from a constant) is

(1.2.37) $$\Phi(u) = \int_\Omega T\sqrt{EG - F^2} - u_i f_i] \, d\Omega - \int_\Gamma u_i g_i \, ds,$$

where $f = (f_1, f_2, f_3)$ is the vector of the surface force related to the unit volume in Ω and $g = (g_1, g_2, g_3)$ is the tension vector on Γ introduced as before; a priori we do not assume any condition (1.2.18) or (1.2.19). If on $\partial\Omega \setminus \Gamma u = u^0$ is prescribed (for example by the initial position), from the variation of (1.2.37) on $O \subset \Omega$, we get for the tension vector

(1.2.38)
$$\sigma = T(EG - F)^{-1/2}\left[G\frac{\partial y}{\partial x_1}\nu_1 + E\frac{\partial y}{\partial x_2}\nu_2 - F\left(\frac{\partial y}{\partial x_2}\nu_1 + \frac{\partial y}{\partial x_1}\nu_2\right)\right].$$

1.2.39. EXERCISE. Prove that for every $O \subset \Omega$, the conditions of total equilibrium

(1.2.40)
$$\int_O f\,dx + \int_{\partial O}\sigma\,ds = 0, \qquad \int_O (y \times f)\,dO + \int_{\partial O}(y \times \sigma)\,ds = 0$$

are satisfied.

It follows easily from (1.2.38) that the conditions (1.2.18) and (1.2.19) are satisfied automatically.

First, as we shall see later, the functional (1.2.37) is of the $[W^{1,1}(\Omega)]^3$ structure or—we can say—of minimal surface type and we shall not study such functionals in these lecture notes. We refer to E. GIUSTI [17]; in general, such a study is very delicate.

1.2.41. EXAMPLE. We shall consider some basic ideas of elasticity; some more modern approach to finite elasticity will be discussed later. From the literature, see for example K. WASHIZU [18], J. NEČAS, I. HLAVÁČEK [19].

Let us consider an elastic body, represented by a bounded domain $\Omega \subset \mathbb{R}^3$. After the deformation, Ω will be transformed into Ω' and we suppose that the mapping $x \mapsto y(x) = x + u(x)$ is a smooth enough diffeomorphism of Ω onto Ω'. The first reasonable hypothesis (which is often neglected) is the condition of the impenetrability of material:

(1.2.42)
$$\det\frac{\partial y_i}{\partial x_j} > 0.$$

The reader can see, for example in [18] or [19], that the deformation is characterized by the Almansi strain tensor

(1.2.43)
$$\varepsilon_{ij} \overset{\text{def}}{=} \frac{1}{2}\left[\frac{\partial u_i}{\partial x_j} + \frac{\partial u_j}{\partial x_i} + \frac{\partial u_k}{\partial x_i}\frac{\partial u_k}{\partial x_j}\right].$$

The main idea how to see it follows classically from the consideration of two points x, $x + \Delta x$ before and $y(x)$, $y(x + \Delta x)$ after the deformation. The deformation is characterized by a change of the distance of two "infinitely close points", i.e., we consider the expression

(1.2.44)
$$\phi(t) = |u(x + t\,\Delta x) + t\,\Delta x - u(x)|^2 - t^2|\Delta x|^2 \doteq 2 \cdot \varepsilon_{ij}\,\Delta x_i\,\Delta x_j.$$

A tension vector σ is introduced in the body Ω' after the deformation: let $O \subset \Omega$, ∂O be smooth enough. For simplicity we write $y(x) = x'$. Let $x' \in \partial O'$

and ν be the outer normal to ∂O at the point x. So $\sigma = \sigma(x, \nu)$ and it is a tension vector related to the unit surface on ∂O. This vector expresses the tension at the point x' of $\Omega' \setminus O'$ onto O'.

The body tension is a vector $F = F(x)$ related to the unit volume in Ω.

The equilibrium conditions for forces and momentum on every $O' \subset \Omega'$ are the basis for our considerations:

$$\int_O F(x)\,\mathrm{d}x + \int_{\partial O} \sigma(x, \nu(x))\,\mathrm{d}S = 0, \tag{1.2.45}$$

$$\int_O (y \times F(x))\,\mathrm{d}x + \int_{\partial O} (y \times \sigma(x, \nu(x)))\,\mathrm{d}S = 0. \tag{1.2.46}$$

If e_j are the basis vectors of the Cartesian coordinate system we define the vectors

$$\sigma^j(x) \overset{\text{def}}{=} \sigma(x, e_j). \tag{1.2.47}$$

We get from (1.2.45) that $\sigma(x, -\nu) = -\sigma(x, \nu)$ and taking a pyramid from Figure 1.2.48 for O, (1.2.45) implies that if the height of the pyramid tends

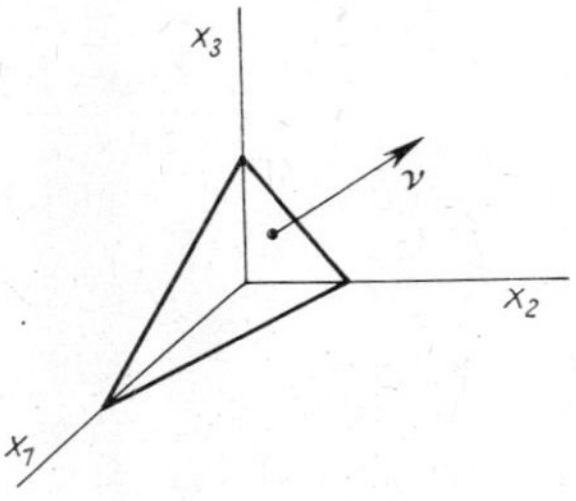

Fig. 1.2.48

to zero (for details see [18] or [19]) then

$$\sigma(x, \nu) = \sigma^i(x)\,\nu_i. \tag{1.2.49}$$

Once more, (1.2.45) gives the equation of equilibrium

$$\frac{\partial \sigma^i}{\partial x_i} + F = 0 \quad \text{in } \Omega \tag{1.2.50}$$

and σ_i^j is called the Piola stress tensor. Let us introduce the contravariant coordinates of σ^j by

$$\sigma^i \overset{\text{def}}{=} \sigma^{ij} \frac{\partial y}{\partial x_j}. \tag{1.2.51}$$

Then σ^{ij} is called the Kirchhoff stress tensor. If we apply (1.2.46), we get

$$\sigma^{ij} = \sigma^{ji} \tag{1.2.52}$$

(for details see [18] or [19]; we use (1.2.45), (1.2.46) and shrink O to a point x regularly). We shall see later that a reasonable Hook's law in finite elasticity is

$$\sigma^{ij} = \frac{\partial A}{\partial \varepsilon_{ij}}, \tag{1.2.53}$$

where $A = A(x, \varepsilon)$ is a function invariant with respect to the replacement of ε_{ij} by ε_{ji}. A simple substitution of (1.2.53) into (1.2.50) gives the basic nonlinear system of finite elasticity

(1.2.54) $$\frac{\partial}{\partial x_i}\left[\frac{\partial A}{\partial \varepsilon_{ik}} + \frac{\partial A}{\partial \varepsilon_{ij}}\frac{\partial u_k}{\partial x_j}\right] + F_k = 0, \quad k = 1, 2, 3.$$

Let us mention that the body force $F = F(x)$ does not depend on the solution u. Let $\Gamma \subset \partial\Omega$ and let g be defined for $x \in \Gamma$, $g = g(x)$. The general problem of finite elasticity is to solve (1.2.54) with the conditions on the boundary:

(1.2.55) $$u = u^0 \quad \text{on } \partial\Omega \setminus \Gamma,$$

(1.2.56) $$\sigma^i \nu_i = g \quad \text{on } \Gamma;$$

of course $\sigma^i \nu_i = \sigma^{ij}\dfrac{\partial y}{\partial x_j}\nu_i = \dfrac{\partial A}{\partial \varepsilon_{ij}}\dfrac{\partial y}{\partial x_j}\nu_i$. We get easily that (1.2.54) is the Euler's equation for the functional of the total potential energy

(1.2.57) $$\Phi(u) = \int_\Omega A(x, \varepsilon(u))\, \mathrm{d}x - \int_\Omega F_i u_i\, \mathrm{d}x - \int_\Gamma g_i u_i\, \mathrm{d}S$$

and that the condition (1.2.56) is also a consequence of (1.2.57): if $\Phi(u)$ attains its minimum at the point u satisfying (1.2.56), then $D\Phi(u, h) = 0\ \forall h = 0$ on $\partial\Omega \setminus \Gamma$, which implies (1.2.54) and (1.2.56).

The reader sees from what spaces is the solution to be taken and how to choose the domain Ω in order that everything what we mentioned may be correct; for example if A is smooth enough, and the data and Ω as well, it is sufficient to take $u \in [C^2(\bar{\Omega})]^3$.

The calculus of variations for the potential (1.2.57) is a very difficult question and there are only relatively recent results for a class of functionals, see J. Ball [20]. If $\Gamma = \emptyset$, some global results, concerning the existence of classical solutions, were obtained by the author, see J. Nečas [21], [22] and Chapter 6.

Let us close the introduction to finite elasticity by a formula the proof of which we leave to the reader: for Φ from (1.2.57) it is

(1.2.58) $$D^2\Phi(u, v, v) = \int_\Omega \left[\frac{\partial^2 A}{\partial \varepsilon_{ij}\,\partial \varepsilon_{kl}}\,\delta\varepsilon_{ij}\,\delta\varepsilon_{kl} + \sigma^{ij}\frac{\partial v_k}{\partial x_i}\frac{\partial v_k}{\partial x_j}\right]\mathrm{d}x.$$

If we suppose $\dfrac{\partial^2 A}{\partial \varepsilon_{ij}\partial \varepsilon_{kl}}\eta_{ij}\eta_{kl} \geqq c|\eta|^2$, it can still happen, because of the term $\sigma^{ij}\dfrac{\partial v_k}{\partial x_i}\dfrac{\partial v_k}{\partial x_j}$, that $D^2\Phi(u, v, v)$ is not $\geqq 0$. So, in general, the functional (1.2.58) is not convex.

1.2.59. EXAMPLE. Hook's law (1.2.53) contains two nonlinearities: the geometrical one, caused by the quadratic term in (1.2.43), and the physical one, caused by the nonlinearity of $\dfrac{\partial A}{\partial \varepsilon_{ij}}$. In linear or nonlinear elasticity, the geometrical nonlinearity is canceled and the strain tensor of small deformation is defined as

(1.2.59)′ $$e_{ij} \overset{\text{def}}{=} \frac{1}{2}\left(\frac{\partial u_i}{\partial x_j} + \frac{\partial u_j}{\partial x_i}\right).$$

The second geometrical linearization consists in introducing the Cauchy stress tensor τ_{ij} on Ω with all references to Ω: as before we introduce the vector of the tension $\tau(x, \nu)$, that means the action of $\Omega \setminus O$ on O at the point $x \in \partial O$ with the outer normal ν to ∂O. Defining, as before,

(1.2.60) $$\tau_i(x) \overset{\text{def}}{=} \tau(x, e_i),$$

we get the Cauchy stress tensor $\tau_{ij} = \tau_j(x, e_i)$. Introducing the vector of body forces related to Ω, we suppose for all $O \subset \Omega$:

(1.2.61) $$\int_O F(x)\,\mathrm{d}x + \int_{\partial O} \tau(x, \nu)\,\mathrm{d}S = 0,$$

(1.2.62) $$\int_O (x \times F(x))\,\mathrm{d}x + \int_{\partial O} (x \times \tau(x, \nu))\,\mathrm{d}S = 0.$$

The relations (1.2.61), (1.2.62) are equivalent (as before) to

(1.2.63) $$\tau(x, \nu) = \tau_i(x)\,\nu_i,$$

(1.2.64) $$\tau_{ij} = \tau_{ji}, \frac{\partial \tau_{ij}}{\partial x_j} + F_i = 0.$$

Hooke's law is supposed in the form

(1.2.65) $$\tau_{ij} = \frac{\partial A}{\partial e_{ij}},$$

the basic system of elasticity is

(1.2.66) $$\frac{\partial}{\partial x_j}\left(\frac{\partial A}{\partial e_{ij}}\right) + F_i = 0, \quad i = 1, 2, 3$$

and the basic boundary conditions are

(1.2.67) $$\tau(x, \nu) = g(x) \quad \text{on } \Gamma \subset \partial\Omega,$$

(1.2.68) $$u = u^0 \quad \text{on} \quad \partial\Omega \setminus \Gamma.$$

We suppose

(1.2.69) $$\frac{\partial^2 A}{\partial e_{ij}\,\partial e_{kl}}\eta_{ij}\eta_{kl} \geqq c|\eta|^2, \qquad \eta_{ij} = \eta_{ji}$$

and we get easily, that (1.2.66) and (1.2.67) follow from the variation of the functional of the total potential energy

(1.2.70) $$\Phi(u) = \int_\Omega A(x, e(u))\,\mathrm{d}x - \int_\Omega u_i F_i\,\mathrm{d}x - \int_\Gamma u_i g_i\,\mathrm{d}S.$$

Chapter 2

Sobolev and Morrey-Campanato spaces

2.1. Definition of the Sobolev spaces

The reader can consult the lecture notes of this series by H. TRIEBEL [23], [24]; from further references let us mention the basic monograph of S. L. SOBOLEV [9]. We shall adhere to the monographs by J. NEČAS [25] and by A. KUFNER, O. JOHN, and S. FUČÍK [26]. From other literature, let us mention the books by O. V. BĚSOV, V. P. IL'IN, S. M. NIKOL'SKIĬ [27] and R. A. ADAMS [28].

Let Ω be a domain, $\Omega \subset \mathbb{R}^n$, where $\mathbb{R}^n$ is an n-dimensional Euclidean space, and let $1 \leqq p \leqq \infty$. Unless otherwise mentioned, we shall be concerned with real functions. Let us denote by $\mathscr{D}(\Omega)$ the linear space of infinitely differentiable functions with a compact support ($\operatorname{supp} \phi = \overline{\{x \in \mathbb{R}^n, \phi(x) \neq 0\}}$).

2.1.1. DEFINITION. Let $1 \leqq p \leqq \infty$. The space $W^{k,p}(\Omega)$ is the subspace of $L^p(\Omega)$-functions u for which there exist $\omega_\alpha \in L^p(\Omega)$, $\alpha = (\alpha_1, \alpha_2, \ldots, \alpha_n)$, $\alpha_i \geqq 0$, $|\alpha| = \alpha_1 + \alpha_2 + \ldots + \alpha_n$, $1 \leqq |\alpha| \leqq k$, such that $\forall \phi \in \mathscr{D}(\Omega)$

(2.1.2) $$\int_\Omega D^\alpha \phi u \, dx = (-1)^{|\alpha|} \int_\Omega \phi \omega_\alpha \, dx ;$$

here $D^\alpha = \dfrac{\partial^{|\alpha|}}{\partial x_1^{\alpha_1} \ldots \partial x_n^{\alpha_n}}$. We put for $1 \leqq p < \infty$

(2.1.3) $$\begin{aligned} \|u\|_{W^{k,p}(\Omega)} &= \|u\|_{W^{k,p}} = \|u\|_{k,p} \\ &= \Big(\int_\Omega \Big[\sum_{1 \leqq |\alpha| \leqq k} |\omega_\alpha|^p \, dx + |u|^p \Big] dx \Big)^{1/p} \end{aligned}$$

and for $p = \infty$

(2.1.4) $$\|u\|_{k,\infty} = \sup_{x \in \Omega} \operatorname{ess} |u(x)| + \sum_{1 \leqq |\alpha| \leqq k} \sup_{x \in \Omega} \operatorname{ess} |\omega_\alpha(x)| .$$

Let us remark that, unless otherwise mentioned, we have this definition in mind in the sequel. We shall denote ω_α by $D^\alpha u$ in what follows.

The reader sees immediately that $W^{k,p}(\Omega)$ is a Banach space; we use the well known fact, that if $f \in L^1_{\mathrm{loc}}(\Omega)$ ($f \in L^1_{\mathrm{loc}}(\Omega)$ if $\forall K$, a compact set from Ω, $f \in L^1(K)$) and $\int_\Omega f\phi \, dx = 0 \ \forall \phi \in \mathscr{D}(\Omega)$, then $f = 0$ almost everywhere.

2.1.5. DEFINITION. Let $\Omega \subset \mathbb{R}^n$ be a bounded domain. Let $1 \leqq p < \infty$. We introduce $W^{k,p}(\Omega) = \overline{C^k(\bar{\Omega})}$, i.e. as the closure in the norm (2.1.3).

Let $f \in L^1_{\mathrm{loc}}(\Omega)$. We say that f *is absolutely continuous on the straight line parallel to the axis* x_i if it is absolutely continuous on closed intervals contained in $P \cap \Omega$. We say that f *is absolutely continuous on almost every parallel to the axis* x_i, if the set of the intersections of the parallels P to the axis x_i with the hyperplane $x_i = 0$, for which $P \cap \Omega \neq \emptyset$ and for which the function f is not absolutely continuous here, has the zero $(n-1)$-dimensional Lebesgue measure.

2.1.6. DEFINITION. Let $\Omega \subset \mathbb{R}^n$, $1 \leqq p \leqq \infty$. The function $f \in L^p(\Omega)$ is from $W^{1,p}(\Omega)$ if it can be changed on a set of zero measure in such a way, that it is absolutely continuous on almost all parallels to the axes $x_1, x_2, \ldots, x_n$ and if its classical derivatives are from $L^p(\Omega)$.

2.1.7. PROPOSITION. *Definition* 2.1.5 $\Rightarrow$ *Definition* 2.1.1.

The proof is obvious.

Let us recall the definition of domains with continuous or Lipschitz continuous boundaries.

2.1.8. DEFINITION. Let $\Omega \subset \mathbb{R}^n$ be a bounded domain. We suppose that the boundary of Ω, $\partial\Omega$, can be mapped in m Cartesian systems by functions $a_r(x')$, $x = (x', x_n)$ (we denote the coordinates in all the systems in question in the same way) continuous in $\overline{\Delta_\alpha} \overset{\text{def}}{=} \{x' \in \mathbb{R}^{n-1} \mid |x_j| \leqq \alpha, \ j = 1, 2, \ldots, n-1\}$ in the form $(x', a_r(x'))$, $|x_j| < \alpha$.

$$\text{Put } \Lambda_r \overset{\text{def}}{=} \{x \in \mathbb{R}^n \mid x' \in \Delta_\alpha, \ x_n = a_r(x')\},$$
$$V_r^+ = \{x \in \mathbb{R}^n \mid x' \in \Delta_\alpha, \ a_r(x') < x_n < a_r(x') + \alpha\},$$
$$V_r^- = \{x \in \mathbb{R}^n \mid x' \in \Delta_\alpha, \ a_r(x') - \alpha < x_n < a_r(x')\}.$$

We suppose $V_r^+ \subset \Omega$, $V_r^- \subset \mathbb{R}^n \setminus \Omega$. In this case, $\partial\Omega$ is called *continuous*. If all the functions a_r are Lipschitz continuous in $\overline{\Delta_\alpha}$, i.e. if $|a_r(x') - a_r(y')| \leqq c|x' - y'|$ in $\overline{\Delta_\alpha}$, then $\partial\Omega$ is called *Lipschitz continuous.*

2.1.9. THEOREM. *Let Ω have a continuous boundary. Let* $1 \leqq p < \infty$. *Then Definition* 2.1.1 $\Rightarrow$ 2.1.5.

Proof: Let $u \in W^{k,p}(\Omega)$ according to Definition 2.1.1. Let $V_{m+1} \subset \overline{V_{m+1}} \subset \Omega$ be such a domain, that $\bigcup_{r=1}^{m=1} V_r \supset \bar{\Omega}$, here $V_r = \Lambda_r \cup V_r^+ \cup V_r^-$, $r = 1, 2, \ldots, m$. There exist $\phi_r \in \mathscr{D}(V_r)$, $0 \leqq \phi_r(x) \leqq 1$ such that $\sum_{r=1}^{m+1} \phi_r(x) = 1$ for $x \in \bar{\Omega}$ (for details see [25]). Put $u_r = u\phi_r$. Clearly $u_r \in W^{k,p}(\Omega)$ according to Definition 2.1.1. Put $u_{r\delta}(x) = u_r(x', x_n + \delta)$ for $\delta > 0$ small enough. Because of the continuity in the mean of a function from $L^p(\Omega)$ (if $f \in L^p(\Omega)$, $\Omega \subset \mathbb{R}^n$ is a bounded domain, $1 \leqq p < \infty$, then $\forall \varepsilon > 0$, $\exists \delta > 0$ such that $|h| < \delta \Rightarrow \int_\Omega |f(x+h) - f(x)|^p \, dx < \varepsilon^p$, $f(x+h) = 0$ if $x + h \notin \Omega$; for details see [25]) we have

$$u_{r\delta} \to u_r \quad \text{in} \quad W^{k,p}(\Omega) \quad \text{for } \delta \to 0. \tag{2.1.10}$$

Let $\omega \in \mathscr{D}(R^n)$ and $\int_{\mathbb{R}_n} \omega(z)\, dz = 1$. Put $u_{r\delta} = 0$ for $x \notin V_r^+ \cup \Lambda_r \cup \{x \mid x' \in \Delta_\alpha,$ $a_r(x') - \delta < x_n < a_r(x')\}$ and then

$$u_{r\delta h}(x) = h^{-n} \int_{\mathbb{R}^n} u_{r\delta}(y)\, \omega\left(\frac{y-x}{h}\right) dy. \tag{2.1.11}$$

Actually $u_{r\delta h} \to u_{r\delta}$ in $W^{k,p}(V_r)$, which follows from the continuity in the mean of $u_{r\delta}$, $D^\alpha u_{r\delta}$ in $L^p(V_r)$

$(u_{r\delta h}(x) = \int_{\mathbb{R}^n} u_{r\delta}(x + hz)\, \omega(z)\, dz$, $1 < p < \infty$, for example,

$$\int_{V_r} |u_{r\delta h}(x) - u_{r\delta}(x)|^p \, dx$$
$$\leqq \int_{V_r} dx \left(\int_{\mathbb{R}^n} |\omega(z)|^{\frac{p}{p-1}} dz \right)^{p-1} \int_{\mathbb{R}^n} |u_{r\delta}(x + hz) - u_{r\delta}(x)|^p \, dz$$
$$\leqq c \int_{\mathbb{R}^n} |u_{r\delta}(x + hz) - u_{r\delta}(x)|^p \, dx, \quad \text{for details see [25])}$$

and from

$$D^\alpha u_{r\delta h}(x) = h^{-n} \int_{\mathbb{R}^n} D^\alpha u_{r\delta}(y)\, \omega\left(\frac{y-x}{h}\right) dy. \tag{2.1.12}$$

Since $u_{r\delta h} \in C^\infty(\bar{\Omega})$ and $u_{m+1h} \to u_{m+1}$ in $W^{k,p}(\Omega)$, and $u_{m+1h} \in \mathscr{D}(\Omega)$, we have that $\sum_{r=1}^{m} u_{rh\delta} + u_{m+1h}$ is close to $u = \sum_{r=1}^{m+1} u_r$ in $W^{k,p}(\Omega)$ as we wish.

2.1.13. PROPOSITION. *If $u \in W^{1,p}(\Omega)$ by Definition* 2.1.6, *then $u \in W^{1,p}(\Omega)$ by Definition* 2.1.1 *as well.*

Proof. Put $u = 0$ for $x \notin \Omega$. We have

$$\int_{\mathbb{R}^n} \frac{\partial \phi}{\partial x_i} u \, dx = \int_{\mathbb{R}^{n-1}} dx' \int_{\mathbb{R}^1} \frac{\partial \phi}{\partial x_i} u \, dx_i$$
$$= - \int_{\mathbb{R}^{n-1}} dx' \int_{\mathbb{R}^1} \phi \frac{\partial u}{\partial x_i} dx_i = - \int_{\mathbb{R}^n} \phi \frac{\partial u}{\partial x_i} dx.$$

2.1.14. THEOREM. *Let $1 \leqq p < \infty$. Let $\bar{\Omega} \subset \Omega_1$ and $u \in W^{1,p}(\Omega_1)$ according to Definition* 2.1.1, *then $u \in W^{1,p}(\Omega)$ according to Definition* 2.1.6.

Proof. Let $\Psi \in \mathscr{D}(\Omega_1)$ be such that $\Psi(x) = 1$ on Ω and put $v = u\Psi$. Clearly $v \in W^{1,p}(\mathbb{R}^n)$ and $\operatorname{supp} v \subset \Omega$. With respect to Theorem 2.1.9, there exist

$v_k \in C^1(\bar{B})$, where B is a ball, $\bar{\Omega} \subset B$, such that $v_k \to v$ in $W^{1,p}(B)$. We can suppose that $v_k \to v$ almost everywhere in B. Hence

$$v_k(x', x_n) = \int_{-\infty}^{x_n} \frac{\partial v_k}{\partial x_n}(x', \xi)\, d\xi . \tag{2.1.15}$$

Choosing subsequences step by step (and we denote all of them in the same way), we get v_k with the property that on almost all parallels with the axis x_n,

$$\lim_{k \to \infty} \int_{-\infty}^{\infty} \left| \frac{\partial v_k}{\partial x_n}(x', \xi) - \frac{\partial v}{\partial x_n}(x', \xi) \right| d\xi = 0 \tag{2.1.16}$$

and an analogue for the other derivatives. Let $v^*(x) = \lim_{k \to \infty} v_k(x)$ at those points, where the sequence converges. We have on almost all parallels with the axis x_n that

$$v^*(x', x_n) = \int_{\infty}^{x_n} \frac{\partial v}{\partial x_n}(x', \xi)\, d\xi \tag{2.1.17}$$

and an analogue for other derivatives. Hence $v^* \in W^{1,p}(\Omega_1)$ according to Definition 2.1.6 and because $v^* = u$ almost everywhere in Ω the proof is finished.

2.1.18. LEMMA. *Let $\Omega \subset \mathbb{R}^n$ be a domain and let $T(y) = x$ be a homeomorphism of Ω onto a domain O such that*

$$c_1 |y^1 - y^2| \leqq |T(y^1) - T(y^2)| \leqq c_2 |y^1 - y^2| . \tag{2.1.19}$$

Let C be an open cube, $\bar{C} \subset \Omega$. Then we have for the Lebesgue measure that

$$\operatorname{meas}(C) \leqq c_3 \operatorname{meas}(T(C)). \tag{2.1.20}$$

Proof. Let y^0 be the centre of the cube C. Because $T(\partial C) = \partial T(C)$ it is for $x \in \partial T(C)$

$$|T(y) - T(y^0)| \geqq c_1 |y - y^0| \geqq \frac{c_1}{2} (\operatorname{meas} C)^{1/n}, \tag{2.1.21}$$

thus also

$$\min_{x \in \partial T(C)} |x - T(y^0)| \geqq \frac{c_1}{2} (\operatorname{meas} C)^{1/n} \tag{2.1.22}$$

and the assertion follows.

2.1.23. THEOREM. *Let T be a homeomorphism of Ω onto a domain O, where $\partial\Omega$ is Lipschitz continuous. Let T satisfy* (2.1.19). *Let $u \in W^{1,p}(\Omega)$, $1 \leqq p \leqq \infty$. Then $u \circ T^{-1} \in W^{1,p}(O)$ and*

$$\|u \circ T^{-1}\|_{W^{1,p}(O)} \leqq c \|u\|_{W^{1,p}(\Omega)} . \tag{2.1.24}$$

Proof. First let $p < \infty$. There exist $u_k \in C^1(\bar{\Omega})$ such that $u_k \to u$ in $W^{1,p}(\Omega)$. The functions $u_k \circ T^{-1} \in C(\bar{O})$ and

$$\int_O |u_k \circ T^{-1}|^p\, dy = \lim_{d \to 0} \sum d^n \min_{y \in \bar{C}_i} |(u_k \circ T^{-1})(y)|^p , \tag{2.1.25}$$

where C_i are cubes with the edge d from the partition of $\mathbb{R}^n$ and the sum in (2.1.25) is taken over all the cubes lying in O. But $\sum_i d^n \min_{y \in \overline{C}_i} |u_k \circ T^{-1}(y)|^p \leqq c_1 \sum_i \operatorname{meas} T(C_i) \min_{x \in T(C_i)} |u_k(x)|^p \leqq c_1 \int_\Omega |u_k(x)|^p \, dx$, and we thus get from (2.1.25) that

(2.1.26) $$\int_O |u \circ T^{-1}|^p \, dy \leqq c_1 \int_\Omega |u|^p \, dx .$$

Functions $u_k \circ T^{-1}$ are Lipschitz continuous in O, so $\dfrac{\partial}{\partial y_j}(u_k \circ T^{-1}) = \dfrac{\partial u_k}{\partial x_i} \dfrac{\partial x_i}{\partial y_j}$ and $\left|\dfrac{\partial x_i}{\partial y_j}\right| \leqq c_2$. We thus get that $u_k \circ T^{-1} \in W^{1,p}(O)$ and, as above,

(2.1.27) $$\int \left| \frac{\partial u_k}{\partial x_i} \circ T^{-1} \right|^p dy \leqq c_2 \int_\Omega \left| \frac{\partial u_k}{\partial x_i} \right|^p dx ;$$

hence $k \to \infty$ gives the assertion for $p < \infty$. But it follows from 2.1.14 that $u \in W^{1,\infty}(\Omega)$ iff u is Lipschitz continuous, so the assertion for $p = \infty$ is obvious.

2.2. Imbedding theorems for Sobolev spaces

2.2.1. THEOREM. *Let Ω be a domain with a Lipschitz continuous boundary. Let $1 \leqq p < n$ and $\dfrac{1}{q} = \dfrac{1}{p} - \dfrac{1}{n}$. Then $W^{1,p}(\Omega) \hookrightarrow L^q(\Omega)$, i.e., $W^{1,p}(\Omega) \subset L^q(\Omega)$ and the identity mapping from $W^{1,p}(\Omega)$ to $L^q(\Omega)$ is bounded.*

If we employ the functions ϕ_r from the proof of Theorem 2.1.9, it is sufficient to prove the theorem for the functions $u\phi_r$. We consider the more difficult case of $r \leqq m$. Let us define a mapping from V_r^+ onto the interval $\Delta_\alpha \times (0, \alpha)$ by

(2.2.2) $$y_i = x_i, \quad i = 1, 2, \ldots, n-1,$$
$$y_n = x_n - a_r(x') .$$

Theorem 2.1.23 thus enables us to prove Theorem 2.2.1 for cubes only and, a fortiori, for the functions that identically vanish in a neighbourhood of all the faces, except for one.

We shall first prove a generalization of the Hölder inequality which is due to E. GAGLIARDO, see [25].

2.2.3. LEMMA. *Let $C = (-1, 1)^n$, $C' = (-1, 1)^{n-1}$ and let $f_i(x') = f_i(x_1, x_2, \ldots, x_{i-1}, x_{i+1}, \ldots, x_n) \in L^{n-1}(C')$. Put $f_i(x) = f_i(x')$ in C. Then*

(2.2.4) $$\int_C \prod_{i=1}^n |f_i| \, dx \leqq \prod_{i=1}^n \left(\int_{C'} |f_i|^{n-1} \, dx' \right)^{1/(n-1)} .$$

Proof by induction. If $n = 2$, (2.2.4) follows from the Fubini theorem. Let $n > 2$ and let us suppose that we have proved (2.2.4) for dimension $n - 1$.

We have

$$I = \int_C |f_1 f_2 \dots f_n| \, dx = \int_{C'} |f_1| \, dx' \int_{-1}^{1} |f_2| \dots |f_n| \, dx_1,$$

hence

$$(2.2.5) \quad I \leqq \int_{C'} |f_1| \prod_{i=2}^{n} \left(\int_{-1}^{1} |f_i|^{n-1} \, dx_1 \right)^{\frac{1}{n-1}} dx'$$

$$\leqq \left(\int_{C'} |f_1|^{n-1} \, dx' \right)^{\frac{1}{n-1}} \left(\int_{C'} \left(\prod_{i=2}^{n} \int_{-1}^{1} |f_i|^{n-1} dx_1 \right)^{\frac{1}{n-2}} dx' \right)^{\frac{n-2}{n-1}}.$$

By induction,

$$(2.2.6) \quad \int_{C'} \left(\prod_{i=2}^{n} \int_{-1}^{1} |f_i|^{n-1} \, dx_1 \right)^{\frac{1}{n-2}} dx' \leqq \prod_{i=2}^{n} \left(\int_{C''} \left(\int_{-1}^{1} |f_i|^{n-1} \, dx_1 \right) dx' \right)^{\frac{1}{n-2}}$$

and the result follows.

Proof of Theorem 2.2.1. Consider $p > 1$ first. Let $u \in W^{1,p}(C)$, C be a cube and u as mentioned above. There exist, as in the proof of Theorem 2.1.9, $u_k \in C^1(\bar{C})$ such that $u_k \to u$ in $W^{1,p}(C)$ and $u_k(x', 1) = 0$. Let us consider $|u_k(x)|^{(np-p)/(n-p)}$. This function is absolutely continuous on all the parallels to the axis and almost everywhere on such a parallel it is

$$(2.2.7) \quad \frac{\partial}{\partial x_i} \left(|u_k|^{\frac{np-p}{n-p}} \right) = \frac{np-p}{n-p} |u_k|^{\frac{np-n}{n-p}} \left| \frac{\partial u_k}{\partial x_i} \right|.$$

From (2.2.7) we get

$$(2.2.8) \quad \max_{0 \leqq x_i \leqq 1} |u_k(x)|^{\frac{np-p}{n-p}} \leqq \frac{np-p}{n-p} \int_0^1 |u_k|^{\frac{np-n}{n-p}} \left| \frac{\partial u_k}{\partial x_i} \right| dx_i .$$

Hence

$$\int_{C'} \max_{0 \leqq x_i \leqq 1} |u_k(x)|^{\frac{np-p}{n-p}} dx' \leqq \frac{np-p}{n-p} \left(\int_C |u_k|^{\frac{np-p}{n-p}} dx \right)^{\frac{p-1}{p}} \left(\int_C \left| \frac{\partial u_k}{\partial x_i} \right|^p dx \right)^{\frac{1}{p}}.$$

(2.2.9)

It thus follows from Lemma 2.2.3 and from (2.2.9) that

$$\int_C |u_k(x)|^{\frac{np}{n-p}} dx \leqq \int_C \left(\prod_{i=1}^{n} \max_{0 \leqq x_i \leqq 1} |u_k(x)|^{\frac{p}{n-p}} \right) dx$$

$$(2.2.10) \quad \leqq \prod_{i=1}^{n} \left(\int_{C'} \max_{0 \leqq x_i \leqq 1} |u_k(x)|^{\frac{np-p}{n-p}} dx' \right)^{\frac{1}{n-1}}$$

$$\leqq \left(\frac{np-p}{n-p} \right)^{\frac{n}{n-1}} \left(\int_C |u_k|^{\frac{np}{n-p}} dx \right)^{\frac{np-n}{np-p}} \|u_k\|_{W^{1,p}(C)}^{\frac{n}{n-1}},$$

therefore

$$(2.2.11) \quad \left(\int_C |u_k(x)|^{\frac{np}{n-p}} dx \right)^{\frac{n-p}{np}} \leqq \frac{np-p}{n-p} \|u_k\|_{W^{1,p}(K)}$$

and as $k \to \infty$, we get the result for $p > 1$. If $p = 1$, it is sufficient to pass to the limit $p \to 1$ in (2.2.11).

Let us recall that $C^{k,\mu}(\bar{\Omega})$ is the subspace of such functions from $C^k(\bar{\Omega})$, whose kth derivatives are μ-Hölder continuous.

The norm is introduced as usual in the form

$$\|u\|_{C^{k,\mu}(\bar\Omega)} \overset{\text{def}}{=} \|u\|_{C^k(\bar\Omega)} + \sum_{|\alpha|=k} \sup_{\substack{x \neq y \\ x,y \in \bar\Omega}} \frac{|D^\alpha u(x) - D^\alpha u(y)|}{|x - y|^\mu}. \tag{2.2.12}$$

2.2.13. THEOREM. *Let Ω be a domain with a Lipschitz boundary. Let $p > n$ and $\mu = 1 - \dfrac{n}{p}$. Then $W^{1,p}(\Omega) \hookrightarrow C^{0,\mu}(\bar{\Omega})$.*

Proof. Similarly as in the proof of Theorem 2.2.1, we can confine ourselves to cubes Ω. Because of the density of $C^1(\bar{\Omega})$ in $W^{1,p}(\Omega)$, we can suppose that the function in question $u \in C^1(\bar{\Omega})$. The limit passage will then give the result. So let $x^1, x^2 \in \bar{C}$—the cube $(0, 1)^n$. There exists a cube C_ϱ with edges of length ϱ, homothetic to C, and such that $x^1, x^2 \in \bar{C}_\varrho$ and

$$\varrho \leqq |x^1 - x^2| \leqq \sqrt{n}\varrho. \tag{2.2.14}$$

Let $x \in \bar{C}_\varrho$. We have

$$|u(x) - u(x^i)| = \left| \int_0^1 \frac{\partial u}{\partial x_j}(x^i + t(x - x^i))(x_j - x^i_j)\,\mathrm{d}t \right| \tag{2.2.15}$$

$$\leqq c_1 \varrho \int_0^1 \sum_{j=1}^n \left| \frac{\partial u}{\partial x_j}(x^i + t(x - x^i)) \right| \mathrm{d}t.$$

Now

$$\left| \frac{1}{\varrho^n} \int_{C_\varrho} u(x)\,\mathrm{d}x - u(x^i) \right| \leqq \varrho^{-n} \int_{C_\varrho} |u(x) - u(x^i)|\,\mathrm{d}x \tag{2.2.16}$$

$$\leqq c_2 \varrho^{-n+1} \int_{C_\varrho} \mathrm{d}x \int_0^1 \sum_{j=1}^n \left| \frac{\partial u}{\partial x_j}(x^i + t(x - x^i)) \right| \mathrm{d}t.$$

The substitution $z = x^i + t(x - x^i)$ leads to

$$\varrho^{-n+1} \int_{C_\varrho} \mathrm{d}x \int_0^1 \sum_{j=1}^n \left| \frac{\partial u}{\partial x_j}(x^i + t(x - x^i)) \right| \mathrm{d}t \tag{2.2.17}$$

$$\leqq \varrho^{-n+1} \int_0^1 t^{-n}\,\mathrm{d}t \int_{C_{\varrho t}} \sum_{j=1}^n \left| \frac{\partial u}{\partial x_j}(z) \right| \mathrm{d}z,$$

where $C_{\varrho t} = \{z \mid z = x^i + t(x - x^i),\ x \in C_\varrho\}$. The Hölder inequality gives

$$\int_{C_{\varrho t}} \left| \frac{\partial u}{\partial x_j}(z) \right| \mathrm{d}z \leqq c_3 (\varrho t)^{\frac{np}{p-1}} \|u\|_{W^{1,p}(C)}, \tag{2.2.18}$$

thus we finally get

$$\left| \frac{1}{\varrho^n} \int_{C_\varrho} u(x)\, dx - u(x^i) \right| \leqq c_4 \varrho^\mu \|u\|_{W^{1,p}(C)}, \tag{2.2.19}$$

hence

$$|u(x^1) - u(x^2)| \leqq c_5 \varrho^\mu \|u\|_{W^{1,p}(C)}. \tag{2.2.20}$$

On the other hand, if $|x^1 - x^2| \geqq \frac{1}{2}$, then $\varrho \geqq \frac{1}{2n}$ and (2.2.19) implies

$$|u(x^1)| \leqq c_6 \|u\|_{W^{1,p}(C)}. \tag{2.2.21}$$

2.3. Complete continuity of imbedding

For more details, see [25], [26].

Let us recall the well known fact (see for example [25]):

2.3.1. CRITERION OF COMPACTNESS. *Let Ω be a bounded domain, $1 \leqq p < \infty$. Let $M \subset L^p(\Omega)$. Then $\overline{M}$ is compact in $L^p(\Omega)$ iff*

$$f \in M \Rightarrow \|f\|_{L^p(\Omega)} \leqq c_1 \tag{2.3.2}$$

and $\forall \varepsilon > 0$, $\exists \delta > 0$ such that for $|h| < \delta$

$$f \in M \Rightarrow \int_\Omega |f(x+h) - f(x)|^p\, dx \leqq \varepsilon^p, \tag{2.3.3}$$

where $f(x) = 0$ for $x \notin \Omega$.

2.3.4. THEOREM. *Let Ω be a domain with a Lipschitz boundary. Then the identity mapping from $W^{1,p}(\Omega)$ to $L^p(\Omega)$ is completely continuous.*

Let us remark, that Theorem 2.3.4 is valid only for $\partial\Omega$ continuous, see [25].

Proof of Theorem 2.3.4. We confine ourselves to the case of $\Omega = (0, 1)^n$ for the reasons mentioned before. Let $C_\eta = (\eta, 1 - \eta)^n$. For functions from $C^1(\overline{C})$, we have $v(x', x_n) = v(x', \xi) + \int_\xi^x \frac{\partial v}{\partial x_n}(x', \omega)\, d\omega$, and thus

$$|v(x', x_n)|^p \leqq 2^{p-1} \left(|v(x', \xi)|^p + \int_0^1 \left| \frac{\partial v}{\partial x_n}(x', \omega) \right|^p d\omega \right) \tag{2.3.5}$$

for $1 \leqq p < \infty$.

The integration of (2.3.5) with respect to ξ over the interval (0, 1) gives

$$|v(x', x_n)|^p \leqq 2^{p-1} \left(\int_0^1 |v(x', \xi)|^p\, d\xi + \int_0^1 \left| \frac{\partial v}{\partial x_n}(x', \omega) \right|^p d\omega \right). \tag{2.3.6}$$

It follows from (2.3.6) that

$$\int_{(0,1)^{n-1}} \mathrm{d}x' \int_{1-\eta}^{1} |v(x', x_n)|^p \, \mathrm{d}x_n \leqq 2^{p-1}\eta\|v\|^p_{W^{1,p}(C)}, \tag{2.3.7}$$

hence, in a similar way, it is

$$\int_{C\setminus C_\eta} |v|^p \, \mathrm{d}x \leqq c_1\eta\|v\|^p_{W^{1,p}(C)}. \tag{2.3.8}$$

Let $|z| < \delta$ and δ be so small that $x \in C_\eta \Rightarrow x + z \in C$. For $v \in C^1(\bar{C})$, we have

$$|v(x + z) - v(x)| = \left|\int_0^1 \frac{\partial v}{\partial x_j}(x + tz)\, z_j \, \mathrm{d}t\right|, \tag{2.3.9}$$

hence we get

$$\int_{C\setminus C_\eta} |v(x + z) - v(x)|^p \, \mathrm{d}x \leqq c_2|z|^p \int_{C\setminus C_\eta} \mathrm{d}x \int_0^1 \sum_{j=1}^{n} \left|\frac{\partial v}{\partial x_j}(x + tz)\right|^p \mathrm{d}t. \tag{2.3.10}$$

Using the substitution $x + tz = y$, we obtain

$$\int_{C\setminus C_\eta} \mathrm{d}x \int_0^1 \sum_{j=1}^{n} \left|\frac{\partial v}{\partial x_j}(x + tz)\right|^p \mathrm{d}t \leqq \int_C \sum_{j=1}^{n} \left|\frac{\partial v}{\partial x_j}(y)\right|^p \mathrm{d}y. \tag{2.3.11}$$

Hence (2.3.8), (2.3.10), (2.3.11) give (2.3.3); of course, we must pass from $C^1(\bar{C})$ to the closure of it: $W^{1,p}(C)$. The condition (2.3.2) is obvious. If $p = \infty$, then it follows from Theorem 2.1.14 that $u \in W^{1,\infty}(C) \Leftrightarrow$

$$|u(x + h) - u(x)| \leqq c\|u\|_{W^{1,\infty}(C)}, \tag{2.3.12}$$

and we have that the imbedding $W^{1,\infty} \subsetneq C(\bar{\Omega})$ is completely continuous.

2.3.13. COROLLARY. *Let Ω be a domain with a Lipschitz boundary. Let $p < n$ and $\dfrac{1}{q} > \dfrac{1}{p} - \dfrac{1}{n}$. Then the identity mapping from $W^{1,p}(\Omega)$ to $L^q(\Omega)$ is completely continuous.*

Proof is sufficient again for $\Omega = (0, 1)^n$. Put $\dfrac{1}{q^*} = \dfrac{1}{p} - \dfrac{1}{n}$. For $v \in L^{q^*}(C)$ we have

$$\int_C |v|^q \, \mathrm{d}x = \int_C |v|^{\frac{q^*(q-1)}{q^*-1}} |v|^{\frac{q^*-q}{q^*-1}} \mathrm{d}x \leqq \left(\int_C |v|^{q^*} \mathrm{d}x\right)^{\frac{q-1}{q^*-1}} \left(\int_C |v| \, \mathrm{d}x\right)^{\frac{q^*-q}{q^*-1}}, \tag{2.3.14}$$

so (2.3.2), (2.3.3) follow from Theorems 2.3.1, 2.3.4 and from (2.3.14).

2.4. Traces

For details see [25], [26], [27]. Let us recall that the Lipschitz boundary is rectifiable and, in local coordinates,

(2.4.1) $$dS = \sqrt{1 + \sum_{i=1}^{n-1} \left(\frac{\partial a}{\partial x_i}\right)^2}\, dx',$$

see Definition 2.1.8.

2.4.2. THEOREM. *Let Ω be a domain with a Lipschitz boundary. Then for $u \in C^1(\bar{\Omega})$,*

(2.4.3) $$\int_{\partial\Omega} |u|^p \, dS \leqq c\|u\|^p_{W^{1,p}(\Omega)}, \quad 1 \leqq p \leqq \infty.$$

Proof. First, with respect to Theorem 2.2.13, there is nothing to be proved for $p > n$. So let $p \leqq n$. According to Definition 2.1.8 we see that

(2.4.4) $$\int_{\partial\Omega} |u|^p \, dS \leqq \sum_{i=1}^{m} \int_{\Delta_i} |u|^p \, dS \leqq c_1 \sum_{i=1}^{m} \int_{\Delta_\alpha} |u|^p \, dx' \leqq c_2 \int_{\partial\Omega} |u|^p \, dS,$$

thus using the partition of unity as in the proof of Theorem 2.1.9 and Theorem 2.1.23, we can again confine ourselves to $\Omega = (0, 1)^n$. Thus

(2.4.5) $$u(x', 1) - u(x', \xi) = \int_0^1 \frac{\partial u}{\partial x_n}(x', \eta) \, d\eta$$

and hence

(2.4.6) $$|u(x', 1)|^p \leqq 2^{p-1} \left((|u(x', \xi)|^p + \int_0^1 \left| \frac{\partial u}{\partial x_n}(x', \eta) \right|^p d\eta \right).$$

The integration with respect to $\xi \in (0, 1)$ gives

(2.4.7) $$|u(x', 1)|^p \leqq 2^{p-1} \left(\int_0^1 \left(|u(x', \xi)|^p + \left| \frac{\partial u}{\partial x_n}(x', \xi) \right|^p \right) d\xi \right),$$

while the integration over $C' = (0, 1)^{n-1}$ gives

(2.4.8) $$\int_{C'} |u(x', 1)|^p \, dx' \leqq 2^{p-1} \|u\|^p_{W^{1,p}(C)}.$$

2.4.9. DEFINITION. Let $p \leqq n$. Define $Tru = u|_{\partial\Omega}$ for $u \in C^1(\bar{\Omega})$. By continuity, extend the operator Tru: $C^1(\bar{\Omega}) \to C(\partial\Omega)$ to the operator $W^{1,p}(\Omega) \to L^p(\partial\Omega)$. Thus for $u \in W^{1,p}(\Omega)$, Tru from $L^p(\partial\Omega)$ is its generalized boundary value (of course, $\partial\Omega$ is a Lipschitz boundary) and we denote it simply by u.

2.4.10. THEOREM. *Let Ω be a domain with a Lipschitz boundary. Let $p < n$. Put* $\frac{1}{q} = \frac{1}{p} - \frac{1}{n-1}\frac{p-1}{p}$. *Then*

$$\|u\|_{L^q(\partial\Omega)} \leqq c\|u\|_{W^{1,p}(\Omega)}. \tag{2.4.11}$$

Let $\frac{1}{q^*} > \frac{1}{p} - \frac{1}{n-1}\frac{p-1}{p}$. *Then the imbedding* $W^{1,p}(\Omega) \to L^{q^*}(\partial\Omega)$ *is completely continuous.*

For the proof, see [25].

Let us put $W_0^{1,p}(\Omega) \overset{\text{def}}{=} \{u \in W^{1,p}(\Omega) \mid u = 0 \text{ on } \partial\Omega, 1 \leqq p < \infty\}$.

Clearly $W_0^{1,p}(\Omega)$ is a closed subspace of $W^{1,p}(\Omega)$, the space of traces is thus characterized by the quotient space $W^{1,p}(\Omega)/W_0^{1,p}(\Omega)$.

2.4.12. THEOREM. *Let Ω be a domain with a Lipschitz boundary. Then* $W_0^{1,p}(\Omega) = \overline{\mathcal{D}(\Omega)}$ *(the closure in* $W^{1,p}(\Omega)$*).*

2.4.13. EXERCISE. Prove 2.4.12. Hint: Extend u by zero into $\mathbb{R}^n \setminus \Omega$. Prove that $u \in W^{1,p}(\mathbb{R}^n)$. Use the partition of unity from the proof of Theorem 2.1.9. For $r = 1, 2, \ldots, m$, use the shifts u_{rh} to the interior of Ω. Then use mollifiers.

2.4.14. DEFINITION. Let $\Delta \subset \mathbb{R}^{n-1}$ be a bounded domain. Let $0 < k < 1$, $1 < p < \infty$. $W^{k,p}(\Delta) \subsetneq L^p(\Delta)$ is a subspace of such functions u for which

$$\|u\|'_{W^{k,p}(\Delta)} \overset{\text{def}}{=} \left(\int_\Delta\int_\Delta \frac{|u(x) - u(y)|^p}{|x - y|^{n-1+pk}}\,\mathrm{d}x\,\mathrm{d}y\right)^{1/p} < \infty. \tag{2.4.15}$$

We put

$$\|u\|_{W^{k,p}(\Delta)} = [\|u\|^p_{L^p(\Delta)} + \|u\|'^p_{W^{k,p}(\Delta)}]^{1/p}. \tag{2.4.16}$$

2.4.17. DEFINITION. Let $\partial\Omega$ of Ω be Lipschitz continuous. Let $0 < k < 1$, $1 < p < \infty$. The space $W^{k,p}(\partial\Omega)$ is defined as the subspace of such functions from $L^p(\partial\Omega)$ for which

$$\int_{\partial\Omega}\int_{\partial\Omega} \frac{|u(x) - u(y)|^p}{|x - y|^{n-1+pk}}\,\mathrm{d}S_x\,\mathrm{d}S_y < \infty. \tag{2.4.18}$$

We put

$$\|u\|^p_{W^{k,p}(\partial\Omega)} \overset{\text{def}}{=} \int_{\partial\Omega} |u|^p\,\mathrm{d}S + \int_{\partial\Omega}\int_{\partial\Omega} \frac{|u(x) - u(y)|^p}{|x - y|^{n-1+pk}}\,\mathrm{d}S_x\,\mathrm{d}S_y. \tag{2.4.19}$$

It is proved for example in [25], [26]:

2.4.20. THEOREM. *Let Ω have a Lipschitz boundary. Let $1 < p < \infty$. Then* $W^{1,p}(\Omega)/W_0^{1,p}(\Omega)$ *is isomorphic to* $W^{1-1/p,p}(\partial\Omega)$ *algebraically and topologically.*

Let us remark that it is possible to construct a linear, bounded map from $W^{1-1/p,p}(\partial\Omega)$ to $W^{1,p}(\Omega)$, an extension of the trace to the whole Ω. For details, see [25].

If $p = 1$, we have that $L^1(\partial\Omega)$ is isomorphic to $W^{1,1}(\Omega)/W_0^{1,1}(\Omega)$, see also [25]. If $p = \infty$, u from $W^{1,\infty}(\Omega)$ are Lipschitz continuous functions and the traces are Lipschitz continuous functions as well. So the connection of $W^{1,\infty}(\Omega)$ with the space of traces is obvious.

2.5. Poincaré inequality

2.5.1. THEOREM. *Let Ω be a domain with a continuous boundary. Let $1 \leqq p < \infty$. Then for $u \in W^{1,p}(\Omega)$:*

$$\int_\Omega \left| u(x) - \frac{1}{\operatorname{meas}\Omega} \int_\Omega u(y)\,\mathrm{d}y \right|^p \mathrm{d}x \leqq c \int_\Omega \sum_{i=1}^n \left| \frac{\partial u}{\partial x_i} \right|^p \mathrm{d}x. \tag{2.5.2}$$

The proof is evidently equivalent to the proof that

$$\int_\Omega |u(x)|^p \,\mathrm{d}x \leqq c \int_\Omega \sum_{i=1}^n \left| \frac{\partial u}{\partial x_i} \right|^p \mathrm{d}x \tag{2.5.3}$$

for the functions from $W^{1,p}(\Omega)$ with a zero mean value.

Let us suppose the contrary. Then there exist $u_k \in W^{1,p}(\Omega)$, $\|u_k\|_{1,p} = 1$ such that

$$\int_\Omega |u_k|^p \,\mathrm{d}x > k \int_\Omega \sum_{i=1}^n \left| \frac{\partial u_k}{\partial x_i} \right|^p \mathrm{d}x. \tag{2.5.4.}$$

It follows from (2.5.4) that

$$\int_\Omega \sum_{i=1}^n \left| \frac{\partial u}{\partial x_i} \right|^p \mathrm{d}x < \frac{1}{k}. \tag{2.5.5}$$

Theorem 2.3.4 implies that we can choose a subsequence $u_{k_l} \to u$ in $L^p(\Omega)$. With regard to (2.5.5), we have $u_{k_l} \to u$ in $W^{1,p}(\Omega)$, $\|u\|_{1,p} = 1$. But it follows also from (2.5.5) that $\dfrac{\partial u}{\partial x_i} = 0$ almost everywhere in Ω; from Theorem 2.1.14 it thus follows that $u = \text{const}$. But u must have also a zero mean value, hence $u = 0$, which contradicts the relation $\|u\|_{1,p} = 1$.

Let us put $B_R(x^0) = \{x \in \mathbb{R}^n \mid |x - x^0| < R\}$, and, for $M \subset \mathbb{R}^n$, meas $M = |M|$.

2.5.6. COROLLARY. *For $u \in W^{1,p}(B_R(x^0))$ we have*

$$\int_{B_R(O)} \left| u(x) - \frac{1}{|B_R(O)|} \int_{B_R(O)} u(y)\,\mathrm{d}y \right|^p \mathrm{d}x \leqq cR^p \int_{B_R(O)} \sum_{i=1}^n \left| \frac{\partial u}{\partial x_i} \right|^p \mathrm{d}x. \tag{2.5.7}$$

We leave the proof to the reader: (2.5.7) is to be reduced by the substitution $x = Rz$ to the unit ball.

2.6. Morrey-Campanato spaces

In this section we use the corresponding part of the books [6] and [26] in a large extent.

2.6.1. DEFINITION. Let $1 \leqq p < \infty$, $0 \leqq \lambda$, and $\Omega \subset \mathbb{R}^n$ be a bounded domain. The space $L^{p,\lambda}(\Omega)$ is the subspace of such functions from $L^p(\Omega)$ for which

$$\|f\|_{L^{p,\lambda}} \overset{\text{def}}{=} \Big\{ \sup_{\varrho > 0,\, x^0 \in \Omega} \varrho^{-\lambda} \int_{B_\varrho(x^0) \cap \Omega} |f|^p \,\mathrm{d}x \Big\}^{\frac{1}{p}} < \infty. \tag{2.6.2}$$

2.6.3. THEOREM. *The space $L^{p,n}(\Omega)$ is isomorphic to $L^\infty(\Omega)$ and*

$$\|f\|_{L^{p,n}} = |B_1(O)|^{1/p}\, \|f\|_{L^\infty}. \tag{2.6.4}$$

Proof. Let $f \in L^\infty(\Omega)$. Then $\int_{\Omega \cap B_\varrho(x^0)} |f|^p \,\mathrm{d}x \leqq \|f\|^p_{L^\infty(\Omega)} |B_1(x^0)|\, \varrho^n$. If $f \in L^{p,n}(\Omega)$ then we have

$$f(x^0) = \lim_{\varrho \to 0} \frac{1}{|B_1(O)|\varrho^n} \int_{\Omega \cap B_\varrho(x^0)} f(x) \,\mathrm{d}x, \tag{2.6.5}$$

for almost all $x^0 \in \Omega$; hence

$$|f(x^0)| \leqq \sup_{\varrho > 0} |B_1(O)|^{-1} \varrho^{-n} \Big(\int_{\Omega \cap B_\varrho(x^0)} |f(x)|^p \,\mathrm{d}x \Big)^{\frac{1}{p}} |B_\varrho(x^0)|^{1 - \frac{1}{p}}, \tag{2.6.6}$$

almost everywhere and thus

$$\|f\|_{L^\infty(\Omega)} \leqq |B_1(x^0)|^{-\frac{1}{p}} \|f\|_{L^{p,n}(\Omega)}.$$

Let us remark a trivial fact, that the only function from $L^{p,\lambda}$ with $\lambda > n$ is zero. Let us write $\Omega_\varrho(x^0) \overset{\text{def}}{=} \Omega \cap B_\varrho(x^0)$ and put

$$u_{x^0,\varrho} = \frac{1}{|\Omega_\varrho(x^0)|} \int_{\Omega_\varrho(x^0)} u(x) \,\mathrm{d}x.$$

2.6.7. DEFINITION. Let $1 \leqq p < \infty$, $0 \leqq \lambda$, and $\Omega \subset \mathbb{R}^n$ be a bounded domain. The space $\mathscr{L}^{p,\lambda}(\Omega)$ is the subspace of such functions from $L^p(\Omega)$, for which

$$[f]_{\mathscr{L}^{p,\lambda}(\Omega)} \overset{\text{def}}{=} \Big(\sup_{\varrho > 0,\, x^0 \in \Omega} \varrho^{-\lambda} \int_{\Omega_\varrho(x^0)} |u - u_{x^0,\varrho}|^p \,\mathrm{d}x \Big)^{\frac{1}{p}} < \infty. \tag{2.6.8}$$

We put

$$\|f\|_{\mathscr{L}^{p,\lambda}(\Omega)} = \|f\|_{L^p(\Omega)} + [f]_{\mathscr{L}^{p,\lambda}(\Omega)}. \tag{2.6.9}$$

Obviously $L^{p,\lambda} \subsetneq \mathscr{L}^{p,\lambda}$ for $0 \leqq \lambda \leqq n$.

2.6.10. THEOREM. *Let Ω have a Lipschitz boundary. Then for $0 \leqq \lambda < n$ the spaces $L^{p,\lambda}(\Omega)$ and $\mathscr{L}^{p,\lambda}(\Omega)$ are isomorphic.*

Proof. For $u \in \mathscr{L}^{p,\lambda}(\Omega)$ we have

$$\varrho^{-\lambda} \int_{\Omega_\varrho(x^0)} |u|^p \,\mathrm{d}x \leqq 2^{p-1} \Big(\varrho^{-\lambda} \int_{\Omega_\varrho(x^0)} |u - u_{x^0,\varrho}|^p \,\mathrm{d}x + |B_1(O)|\, \varrho^{-\lambda+n} |u_{x^0,\varrho}|^p \Big). \tag{2.6.11}$$

Let us confine ourselves to such ϱ (otherwise the situation is obvious) that $R_{h+1} = \varrho = R\,2^{-(1+h)}$, where diam $\Omega < R \leqq 2$ diam Ω. We assert that

$$|u_{x^0,R} - u_{x^0,R_{h+1}}| \leqq c_1 R_{h+1}^{\frac{\lambda-n}{p}} [u]_{\mathscr{L}^{p,\lambda}}. \tag{2.6.12}$$

Let us first suppose that (2.6.12) is proved. Then

$$\begin{aligned}|u_{x^0,\varrho}|^p &\leqq 2^{p-1}(|u_{x^0,R}|^p + |u_{x^0,R} - u_{x^0,\varrho}|^p)\\ &\leqq 2^{p-1}(|u_{x^0,R}|^p + c_1^p \varrho^{\lambda-n}[u]^p_{\mathscr{L}^{p,\lambda}}),\end{aligned} \tag{2.6.13}$$

hence

$$\varrho^{-\lambda+n}|u_{x^0,\varrho}|^p \leqq 2^{p-1}(\varrho^{n-\lambda}|u_{x^0,R}|^p + c_1^p[u]^p_{\mathscr{L}^{p,\lambda}}),$$

and we get $\mathscr{L}^{p,\lambda} \subsetneq L^{p,\lambda}$ from (2.6.11). Thus let $0 < r < R$. We have

$$|u_{x^0,R} - u_{x^0,r}|^p \leqq 2^{p-1}(|u(x) - u_{x^0,R}|^p + |u(x) - u_{x^0,r}|^p) \tag{2.6.14}$$

and the integration over $\Omega_r(x^0)$ gives

$$\begin{aligned}|u_{x^0,R} - u_{x^0,r}|^p \leqq 2^{p-1}c_2 r^{-n}\Big(&\int_{\Omega_R(x^0)} |u(x) - u_{x^0,R}|^p\, dx\\ &+ \int_{\Omega_r(x^0)} |u(x) - u_{x^0,r}|^p\, dx\Big).\end{aligned} \tag{2.6.15}$$

From this we have

$$|u_{x^0,R} - u_{x^0,r}| \leqq c_3 r^{-\frac{n}{p}} R^{\frac{\lambda}{p}} [u]_{\mathscr{L}^{p,\lambda}}. \tag{2.6.16}$$

If we put $R_i = R\,2^{-i}$, $i = 0, 1, \ldots$, then (2.6.16) gives

$$|u_{x^0,R_i} - u_{x^0,R_{i+1}}| \leqq c_3 R^{\frac{\lambda-n}{p}} 2^{i\frac{n-\lambda}{p} + \frac{n}{p}} [u]_{\mathscr{L}^{p,\lambda}}. \tag{2.6.17}$$

Therefore

$$|u_{x^0,R} - u_{x^0,R_{h+1}}| \leqq c_4 R_{h+1}^{\frac{\lambda-n}{p}} [u]_{\mathscr{L}^{p,\lambda}}. \tag{2.6.18}$$

2.6.19. REMARK. The reader sees easily that, for $\lambda = n$, the space $\mathscr{L}^{p,n}$ is larger than $L^{p,n}$. For example $\log x \in \mathscr{L}^{1,1}((0, 1))$.

2.6.20. THEOREM. *Let Ω be a domain with a Lipschitz boundary. Let $n < \lambda \leqq n + p$. Then $\mathscr{L}^{p,\lambda}(\Omega)$ is isomorphic to $C^{0,\alpha}(\bar{\Omega})$, $\alpha = \dfrac{\lambda - n}{p}$.*

Proof. Let us first suppose that $u \in C^{0,\alpha}(\bar{\Omega})$. For $x \in \Omega_\varrho(x^0)$ we have $|u(x) - u_{x^0,\varrho}| = |\frac{1}{|\Omega_\varrho(x^0)|} \int_{\Omega_\varrho(x^0)} (u(x) - u(y))\, dy| \leqq \|u\|_{C^{0,\alpha}(\Omega)} 2^\alpha \varrho^\alpha$, hence $\varrho^{-\lambda} \int_{\Omega_\varrho(x^0)} |u(x) - u_{x^0,\varrho}|^p\, dx \leqq c_1 \varrho^{n-\lambda+\alpha p} \|u\|^p_{C^{0,\alpha}(\Omega)}$, and thus

$$[u]_{\mathscr{L}^{p,\lambda}(\Omega)} \leqq c_2 \|u\|_{C^{0,\alpha}(\Omega)}. \tag{2.6.20}$$

Let $u \in \mathscr{L}^{p,\lambda}(\Omega)$ and $R > 0$, and put $R_i = R\,2^{-i}$, $i = 0, 1, \ldots$, as above. Instead of (2.6.18) we get

$$|u_{x^0,R} - u_{x^0,R_h}| \leqq c_3 R^{\frac{\lambda-n}{p}} [u]_{\mathscr{L}^{p,\lambda}}$$

it thus follows that

$$|u_{x^0,R_h} - u_{x^0,R_h}| \leqq c_3 R_h^{\frac{\lambda-n}{p}} [u]_{\mathscr{L}^{p,\lambda}}. \tag{2.6.21}$$

Hence the sequence u_{x_0, R_h} is a Cauchy sequence and we denote its limit by $u(x^0)$, which has, of course, the usual meaning almost everywhere. It is clear that $u(x^0)$ is independent of the choice of R. (2.6.21) implies

(2.6.22) $$|u_{x,R} - u(x)| \leqq c_3 R^{\frac{\lambda-n}{p}} [u]_{\mathscr{L}^{p,\lambda}}.$$

Let $x, y \in \bar{\Omega}$, $R = |x - y|$. We have

(2.6.23) $$|u(x) - u(y)| \leqq |u_{x,2R} - u(x)| + |u_{x,2R} - u_{y,2R}| + |u_{y,2R} - u(y)|.$$

Put $S = \Omega_{2R}(x) \cap \Omega_{2R}(y)$. We further have

(2.6.24) $$|u_{x,2R} - u_{y,2R}| \leqq \frac{1}{|S|} \int_{\Omega_{2R}(x)} |u_{x,2R} - u(z)|\, dz$$

$$+ \frac{1}{|S|} \int_{\Omega_{2R}(y)} |u_{y,2R} - u(z)|\, dz \leqq c_4 R^{-n+n\left(1-\frac{1}{p}\right)+\frac{\lambda}{p}} [u]_{\mathscr{L}^{p,\lambda}}.$$

Hence (2.6.23)–(2.6.24) imply

(2.6.25) $$|u(x) - u(y)| \leqq c_5 |x - y|^{\frac{\lambda-n}{p}} [u]_{\mathscr{L}^{p,\lambda}}.$$

If we choose $\operatorname{diam} \Omega < R \leqq 2 \operatorname{diam} \Omega$ in (2.6.22) we get

(2.6.26) $$\max_{x \in \bar{\Omega}} |u(x)| \leqq c_6 \|u\|_{\mathscr{L}^{p,\lambda}}.$$

Chapter 3

Existence of weak solutions to boundary value problems for nonlinear second order elliptic systems

A very good book devoted to the theory of monotone operators is the book by D. PASCALI, S. SBURLAN [29]. From other basic monographs, see H. BRÉZIS [30], H. GAJEWSKI, K. GRÖGER, K. ZACHARIAS [31], S. FUČÍK, J. NEČAS, J. SOUČEK [7], E. ZEIDLER [8], S. FUČÍK, A. KUFNER [5], J.L. LIONS [3], S. FUČÍK, J. NEČAS, J. SOUČEK, V. SOUČEK [32], I.V. SKRYPNIK [33], CH.B. MORREY [2], and others.

As far as the papers are concerned, we mention only the fundamental paper by H. BRÉZIS, L. NIRENBERG [34] dealing with the result formulated below by E. LANDESMAN and A.C. LAZER; of course, in the individual sections we add some references we shall use.

3.1. Weak solution

We consider a system in the divergence form

$$(3.1.1) \qquad -\frac{\partial}{\partial x_i}[a_i^r(x, u, \nabla u)] + a^r(x, u, \nabla u) = -\frac{\partial f_i^r}{\partial x_i} + f^r,$$

mentioned in the introduction.

We thus consider a bounded domain $\Omega \subset \mathbb{R}^n$, $n \geqq 2$, with a Lipschitz boundary $\partial\Omega$ and we suppose in any case that the functions f_i^r, f^r are measurable in Ω and that the functions $a_i^r(x, \xi, \eta)$ for $\xi \in \mathbb{R}^m$, $\eta \in \mathbb{R}^{mn}$ satisfy the so-called *Carathéodory conditions*:

3.1.2. DEFINITION. $\forall \xi \in \mathbb{R}^m$, $|\xi| < a$, $\forall \eta \in \mathbb{R}^{mn}$, $|\eta| < b$, the function $a_i^r(x, \xi, \eta)$ is measurable in Ω and, for almost all x from Ω, the function $a_i^r(x, \xi, \eta)$ is continuous in ξ, η in the domain of definition $|\xi| < a$, $|\eta| < b$.

Usually, one supposes $a = b = \infty$ and some growth condition. A typical growth condition is the polynomial growth: there exists $1 < p < \infty$ such that

$$(3.1.3) \qquad |a_i^r(x, \xi, \eta)| \leqq c(1 + |\xi| + |\eta|)^{p-1} + g_i^r(x),$$

$$g_i^r \in L^{p^*}(\Omega), \qquad \left(\frac{1}{p^*} + \frac{1}{p} = 1\right).$$

Under the conditions of ellipticity, which follow, (and some supplementary conditions for lower order terms) the existence theory is essentially the same for $1 < p < \infty$; the case of $p = 1$ is more complicated, see for example J. P. GOSSEZ [35] where also the case of the exponential growth is considered.

In this chapter, we confine ourselves to the polynomial growth of the type (3.1.3) and its natural, so-called "controllable", generalization. Besides (3.1.3) we thus suppose the Carathéodory condition for a^r and the growth conditions

$$(3.1.4) \qquad |a^r(x, \xi, \eta)| \leqq g^r(x) + c|\xi|^{p-1} + c|\eta|^{p-1}, \quad g^r \in L^{p^*}(\Omega).$$

Let us suppose that $f_i^r \in L^{p^*}(\Omega)$, $f^r \in L^{p^*}(\Omega)$. Let $\Gamma \subset \partial\Omega$ be an open subset of $\partial\Omega$. Let a_{lk} be a matrix with coefficients from $L^\infty(\Gamma)$. Let the subscript l runs through the set of subscripts $0 \leqq l_1 < l_2 < \ldots < l_{m_1} \leqq m$ and the subscript k through the complementary set of subscripts $0 \leqq k_1 < k_2 < \ldots < k_{m_2} \leqq m$. We do not exclude that some of these sets are empty. We thus define the linear set $V_\Gamma \overset{\text{def}}{=} \{u \in [W^{1,p}(\Omega)]^m \mid u_l = a_{lk}u_k \text{ on } \Gamma\}$. Let $\partial\Omega = \bigcup_{i=1}^{\varkappa} \Gamma_i \cup M$, where $\text{meas}_{n-1} M = 0$; let a_{lk}^i be given (with different sets of indices) on every Γ_i, and let $V = \bigcap_{i=1}^{\varkappa} V_{\Gamma_i}$. We have

$$(3.1.5) \qquad [W_0^{1,p}(\Omega)]^m \subsetneqq V \subsetneqq [W^{1,p}(\Omega)]^m.$$

Let $u^0 \in [W^{1,p}(\Omega)]^m$ and $g \in [L^{p^*}(\partial\Omega)]^m$ be given. We suppose further that functions $A^r(x, \xi)$ satisfying the Carathéodory conditions and the growth condition

$$(3.1.6) \qquad |A^r(x, \xi)| \leqq h^r(x) + c|\xi|^{p-1}, \quad h^r \in L^{p^*}(\partial\Omega)$$

are defined on $\partial\Omega \times \mathbb{R}^m$.

3.1.7. DEFINITION. Let a function $h(x, \lambda)$ satisfy the Carathéodory condition on $\Omega \times \mathbb{R}^1$ and let

$$(3.1.8) \qquad |h(x, \lambda)| \leqq g(x) + c|\lambda|^{p-1}, \quad g \in L^{p^*}(\Omega), \quad 1 \leqq p < \infty.$$

The operator $u \mapsto h(x, u(x))$ is called the Nemyckiĭ operator.

3.1.9. THEOREM. *The Nemyckiĭ operator is a continuous map from $L^p(\Omega)$ to $L^{p^*}(\Omega)$.*

For the proof see for example [7] or [10]; we shall prove the theorem provided that the function $h(x, \lambda)$ is continuous on $\bar{\Omega} \times R^1$ and $1 < p < \infty$. First let $u \in L^p(\Omega)$. There exists $u_k \in C(\bar{\Omega})$ such that $u_k \to u$ in $L^p(\Omega)$ and almost everywhere in Ω. Of course, $h(x, u_k(x)) \in C(\bar{\Omega})$ and $h(x, u_k(x)) \to h(x, u(x))$ almost everywhere in Ω, hence $h(x, u(x))$ is a measurable function. It follows from (3.1.8) that $h(x, u(x)) \in L^{p^*}(\Omega)$. If $v_k \in L^p(\Omega)$ and $v_k \to u$ in $L^p(\Omega)$ and almost everywhere in Ω, then by the Jegorof theorem $\forall \varepsilon > 0$, $\exists M \subset \Omega$, $|M| < \varepsilon$, such that $v_k \to u$ uniformly in $\Omega \setminus M$. Using (3.1.8), we see that $h(x, v_k(x)) \to h(x, u(x))$ in $L^{p^*}(\Omega)$. (The reader will excuse me, that I shall not use the exact, but uncomfortable notation $h(\cdot, v_k(\cdot)) \to h(\cdot, u(\cdot))$ and so on ... .) If $v_k \to u$ in $L^p(\Omega)$ only, then we can choose a subsequence which tends to u almost everywhere in Ω. Hence, from every subsequence we can choose a subsequence whose limit is $h(x, u(x))$ in $L^{p^*}(\Omega)$, hence the whole $h(x, v_k(x)) \to h(x, u(x))$ in $L^{p^*}(\Omega)$.

The reader himself can, of course, easily generalize Theorem 3.1.9 to the mappings $u \mapsto a_i^r(x, u, \nabla u)$ from $[W^{1,p}(\Omega)]^m$ to $L^{p*}(\Omega)$ and so on.

3.1.10. DEFINITION. The *weak solution* to the system (3.1.1) satisfying the stable boundary condition

$$u - u^0 \in V \tag{3.1.11}$$

is such a function u from $[W^{1,p}(\Omega)]^m$, for which, $\forall v \in V$, it holds

$$\int_\Omega \left[a_i^r(x, u, \nabla u)\frac{\partial v_r}{\partial x_i} + a^r(x, u, \nabla u)\, v_r - f_i^r \frac{\partial v_r}{\partial x_i} - f^r v_r\right] \mathrm{d}x + \int_{\partial\Omega} [A^r(x, u)\, v_r - g^r v_r]\, \mathrm{d}S = 0. \tag{3.1.12}$$

If $V \neq [W_0^{1,2}(\Omega)]^m$ we suppose $f_i^r \equiv 0$.

3.1.13. REMARK. A formal interpretation that follows explains the meaning of 3.1.10. It will follow that for some choice of V the definition of u^0 or g is useless.

3.1.14. INTERPRETATION. Let us suppose that $u \in [C^2(\bar{\Omega})]^m$, $f^r \in C^1(\bar{\Omega})$. The relation (3.1.11) means that we have

$$u_l - a_{lk}^t u_k = u_l^0 - a_{kl}^t u_k^0 \overset{\text{def}}{=} h_l, \tag{3.1.15}$$

on Γ_t, thus we have m_1^t conditions on Γ_t. Let $v \in [W_0^{1,p}(\Omega)]^m$. Then the integration by parts gives

$$0 = \int_\Omega \left\{-\frac{\partial}{\partial x_i}[a_i^r(x, u, \nabla u)] + a^r(x, u, \nabla u) + \frac{\partial f_i^r}{\partial x_i} - f^r\right\} v_r \,\mathrm{d}x = 0, \tag{3.1.16}$$

hence the equations

$$-\frac{\partial}{\partial x_i}[a_i^r(x, u, \nabla u)] + a^r(x, u, \nabla u) = -\frac{\partial f_i^r}{\partial x_i} + f^r \tag{3.1.17}$$

are fulfilled. Let $v \in V$ be such that $v = 0$ on $\partial\Omega \setminus \bigcup_{t=2}^{\varkappa} \Gamma_t$. Green's theorem gives (we suppose $f_i^r = 0$)

$$(3.1.18)\quad 0 = \int_\Omega \left[a_i^r(x, u, \nabla u) \frac{\partial v_r}{\partial x_i} + a^r(x, u, \nabla u)\, v_r - f^r v_r\right] dx$$

$$+ \int_{\Gamma_1} [A^r(x, u)\, v_r - g^r v_r]\, dS$$

$$= \int_{\Gamma_1} [a_i^r(x, u, \nabla u)\, \nu_i + A^r(x, u) - g^r]\, v_r\, dS$$

$$+ \int_\Omega \left\{- \frac{\partial}{\partial x_i} [a_i^r(x, u, \nabla u)] + a^r(x, u, \nabla u) - f^r\right\} v_r\, dx$$

$$= \int_{\Gamma_1} \{a_i^k(x, u, \nabla u)\, \nu_i + A^k(x, u) - g^k + [a_i^l(x, u, \nabla u)\, \nu_i$$

$$+ A^l(x, u) - g^l]\, a_{lk}^1\}\, v_k\, dS.$$

Owing to the "independence" of the functions v_k (for the subscripts $0 \leqq k_1 < k_2 < \ldots < k_{m_2^1} \leqq m$) on Γ_1, we "get" other m_2^1 conditions on Γ_1

$$(3.1.19)\quad a_i^k(x, u, \nabla u)\, \nu_i + A^k(x, u) + [a_i^l(x, u, \nabla u)\, \nu_i + A^l(x, u)]\, a_{lk}^1 = g^k + g^l a_{lk}^1.$$

Since the conditions (3.1.19) contain ∇u and are not stable in the topology of $W^{1,p}(\Omega)$, we call them *unstable*.

The problem with $V = [W_0^{1,p}(\Omega)]^m$ is called the *Dirichlet* problem and the problem with $V = [W^{1,p}(\Omega)]^m$, $A^r(x, u) \equiv 0$ the *Neumann* one, the problem with $V = [W^{1,p}(\Omega)]^m$ and $A^r(x, u) \not\equiv 0$ the *Newton* one, and other problems corresponding to $[W_0^{1,p}(\Omega)]^m \neq V \neq [W^{1,p}(\Omega)^m$ are called *intermediary*; sometimes, if $\partial\Omega = \bigcup_{i=1}^{\varkappa} \Gamma_i$ and $\varkappa \geqq 2$, we speak about combined problems or about mixed problems. The latter term is not very suitable because this notion is usually reserved for evolutionary equations.

Only for information we shall formulate the problem for higher order systems, confining ourselves to the Dirichlet, Neumann or Newton problem without giving precision to the behaviour of all the considered functions.

Thus let the spaces $W^{\varkappa_r,p}(\Omega)$, $r = 1, 2, \ldots, m$, $\varkappa_r \geqq 1$, $1 < p < \infty$ be given and define $W_0 = W_0^{\varkappa_1,p}(\Omega) \times W_0^{\varkappa_2,p}(\Omega) \times \ldots \times W_0^{\varkappa_m,p}(\Omega)$, $W = W^{\varkappa_1,p}(\Omega) \times \times W^{\varkappa_2,p}(\Omega) \times \ldots \times W^{\varkappa_m,p}(\Omega)$. Let $u^0 \in W$, $f_\alpha^r \in L^{p^*}(\Omega)$ for $|\alpha| \leqq \varkappa_r$. Let the functions $a_\alpha^r(x, \xi, \eta)$, $|\alpha| \leqq \varkappa_r$, be given, where $x \in \Omega$, the vector ξ corresponds to all the derivatives of the functions u_s, $s = 1, 2, \ldots, m$, up to the order $\varkappa_s - 1$ and the vector η corresponds to all the derivatives of the functions u_s, $s = 1, 2, \ldots, m$, of order $|\alpha| = \varkappa_s$. Let $A_\alpha^r(x, \xi)$ be given for $x \in \partial\Omega$, $|\alpha| \leqq \varkappa_r - 1$, and the functions $g_\alpha^r \in L^{p^*}(\partial\Omega)$ for $|\alpha| \leqq \varkappa_r - 1$.

3.1.20. DEFINITION. The weak solution to the *Dirichlet problem* for a general elliptic system is $u \in W$ such that

$$(3.1.21)\qquad u - u^0 \in W_0,$$

$$(3.1.22)\qquad \forall v \in W_0 \quad \text{it is} \quad \int_\Omega [a_\alpha^r(x, \delta u, \nabla u)\, D^\alpha v_r - f_\alpha^r D^\alpha v_r]\, dx = 0.$$

The weak solution to the *Neumann* or *Newton* problem is $u \in W$ such that, $\forall v \in W$,

(3.1.23)
$$\int_\Omega [a_\alpha^r(x, \delta u, \nabla u)\, D^\alpha v_r - f_\alpha^r D^\alpha v_r]\, dx + \int_{\partial\Omega} [A_\alpha^r(x, \delta u) - g_\alpha^r]\, D^\alpha v_r\, dS = 0;$$

the summation in the surface integral is taken over $|\alpha| \leqq \varkappa_r - 1$.

We can clearly have $W_0 \subsetneq V \subsetneq W$ and define an intermediary problem. This includes also nonlocal problems.

3.1.24. EXAMPLE. Let $K_{lk}(x, y)$ be a matrix of bounded kernels on $\partial\Omega \times \partial\Omega$. The subscripts and a_{lk} were defined above. Let $V = \{v \in [W^{1,p}(\Omega)]^m \mid v_l(x) = \int_{\partial\Omega} K_{lk}(x, y)\, v_l(y)\, dS_Y + a_{lk}(x)\, v_l(x)\}$. It is also $[W_0^{1,p}(\Omega)]^m \subsetneq V \subsetneq [W^{1,p}(\Omega)]^m$ but V does not correspond to a standard boundary value problem.

One of the most surprising gaps in the theory of elliptic systems is the consideration of only linear constraints among traces. In general, it is an open

3.1.25. PROBLEM. Let a nonlinear constraint

(3.1.26)
$$u_l = B_l(x, u_{k_1}, u_{k_2}, \ldots, u_{k_{m_2}}),$$

$l = l_1, l_2, \ldots, l_{m_1}$, be given on $\partial\Omega$. Let us suppose, for example,

(3.1.27)
$$\left|\frac{\partial B_l}{\partial u_{k_i}}\right| \leqq c.$$

Let $M = \{u \in [W^{1,p}(\Omega)]^m \mid u \text{ satisfies (3.1.26) on } \partial\Omega\}$. Let M_u be the space tangent to M at the point u. We look for $u \in [W^{1,p}(\Omega)]^m \cap M$ such that

(3.1.28) $\forall v \in M_u$,
$$\int_\Omega \left[a_i^r(x, u, \nabla u) \frac{\partial v_r}{\partial x_i} + a^r(x, u, \nabla u)\, v_r - f^r v_r\right] dx + \int_{\partial\Omega} [A^r(x, u) - g^r]\, v_r\, dS = 0.$$

The solution of this problem 3.1.25 is easy if (3.1.28) is an Euler's system, see later.

3.2. Variational approach

Let a function $F(x, \xi, \eta)$ on $\Omega \times \mathbb{R}^m \times \mathbb{R}^{mn}$ be given and let it satisfy the Carathéodory conditions. Let

(3.2.1)
$$|F(x, \xi, \eta)| \leqq f(x) + c|\xi|^p + c|\eta|^p, \quad f \in L^1(\Omega).$$

Let $G(x, \xi)$ satisfy the Carathéodory conditions on $\partial\Omega \times \mathbb{R}^m$ and let the growth conditions

(3.2.2)
$$|G(x, \xi)| \leqq g(x) + c|\xi|^p, \quad g \in L^1(\partial\Omega)$$

be fulfilled.

As before, the space V is given, $u^0 \in [W^{1,p}(\Omega)]^m$, $f_i^r \in L^{p^*}(\Omega)$, $f^r \in L^{p^*}(\Omega)$ (and $f_i^r = 0$ if $V \neq [W_0^{1,p}(\Omega)]^m$), $g^r \in L^{p^*}(\partial\Omega)$; of course, $1 < p < \infty$. Let us consider the functional

(3.2.3) $$\Phi(u) = \int_\Omega \left[F(x, u, \nabla u) - f_i^r \frac{\partial u_r}{\partial x_i} - f^r u_r\right] \mathrm{d}x + \int_{\partial\Omega} [G(x, u) - g^r u_r] \,\mathrm{d}S.$$

3.2.4. THEOREM. *For almost all $x \in \Omega$, let there exist continuous derivatives of $F(x, \xi, \eta)$ in ξ and η and let*

(3.2.5) $$\left|\frac{\partial F}{\partial \xi_r}\right| + \left|\frac{\partial F}{\partial \eta_i^r}\right| \leqq F(x) + c|\xi|^{p-1} + c|\eta|^{p-1},$$

$F \in L^{p^*}(\Omega)$, $r = 1, 2, \ldots, m$, $i = 1, 2, \ldots, n$. For almost all $x \in \partial\Omega$, let the function $G(x, \xi)$ be once continuously differentiable with respect to ξ and let

(3.2.6) $$\left|\frac{\partial G}{\partial \xi_r}\right| \leqq g(x) + c|\xi|^{p-1}, \quad g \in L^{p^*}(\partial\Omega).$$

Then there exists the Gâteaux differential (in fact, the Fréchet differential) of

$$\Phi : D\Phi(u, v) \overset{\text{def}}{=} \frac{\mathrm{d}}{\mathrm{d}t} \Phi(u + tv)\Big|_{t=0} \quad \text{for} \quad v \in [W^{1,p}(\Omega)]^m$$

(automatically we suppose that the Gâteaux differential is a linear, bounded functional):

(3.2.7) $$D\Phi(u, v) = \int_\Omega \left[\frac{\partial F}{\partial \eta_i^r}(x, u, \nabla u)\frac{\partial v_r}{\partial x_i} + \frac{\partial F}{\partial \xi_r}(x, u, \nabla u)\, v_r - f_i^r \frac{\partial v_r}{\partial x_i} - f^r v_r\right] \mathrm{d}x + \int_{\partial\Omega} \left[\frac{\partial G}{\partial \xi_r}(x, u) - g_r\right] v_r \,\mathrm{d}S.$$

The proof follows immediately from the rule for the differentiation of an integral with respect to a parameter: for example, for $|t| \leqq 1$,

$$\left|\frac{\partial F}{\partial \eta_i^r}(x, u + tv, \nabla u + t\,\nabla v)\frac{\partial v_r}{\partial x_i}\right| \leqq F(x) + c_1(|u|^{p-1} + |v|^{p-1} + |\nabla u|^{p-1} + |\nabla v|^{p-1})\,|\nabla v|$$

which is an integrable majorant.

It follows from (3.2.7) that under the conditions of continuous differentiability of the functions a_i^r, a^r, A^r, the necessary condition for the potentiality of the system (3.1.1) is

(3.2.8) $$\frac{\partial a_i^r}{\partial \eta_j^s} = \frac{\partial a_j^s}{\partial \eta_i^r}, \quad \frac{\partial a_i^r}{\partial \xi_s} = \frac{\partial a_s}{\partial \eta_i^r}, \quad \frac{\partial a^r}{\partial \xi_s} = \frac{\partial a^s}{\partial \xi_r}, \quad \frac{\partial A^r}{\partial \xi_s} = \frac{\partial A^s}{\partial \xi_r}.$$

In this case we get, apart from some additive functions,

$$F(x, \xi, \eta) = \int_0^1 [a_i^r(x, t\xi, t\eta)\, \eta_i^r + a^r(x, t\xi, t\eta)\, \xi_r]\, dt, \tag{3.2.9}$$

$$G(x, \xi) = \int_0^1 A^r(x, t\xi)\, \xi_r\, dt. \tag{3.2.10}$$

We also get, apart from some constant,

$$\Phi(u) = \int_0^1 D\Phi(tu, u)\, dt; \tag{3.2.11}$$

this makes sense because $D\Phi(tu, u)$ is continuous in t, in fact, $D\Phi(u, \cdot)$ is continuous from $[W^{1,p}(\Omega)]^m$ to $\{[W^{1,p}(\Omega)]^m\}^*$. See later.

In the general case we have (see J. NEČAS [40]):

3.2.12. THEOREM. *Let* a_i^r, a^r, A^r *satisfy* (3.1.3), (3.1.4), (3.1.6). *Let the conditions* (3.2.8) *be satisfied in the sense of distributions; for example, it is*

$$\int_{R^{m(n+1)}} \left[a_i^r(x, \xi, \eta) \frac{\partial \phi}{\partial \eta_j^s}(\xi, \eta) - a_i^s(x, \xi, \eta) \frac{\partial \phi}{\partial \eta_i^r}(\xi, \eta) \right] d\xi\, d\eta = 0\ \forall\, \phi \in \mathscr{D}(\mathbb{R}^{m(n+1)}). \tag{3.2.13}$$

Then with (3.2.9), (3.2.10), *the expression* (3.1.12) *is the Gâteaux differential of the potential* (3.2.3).

Proof. Let $\omega \in \mathscr{D}(\mathbb{R}^{m(n+1)})$, $\int_{\mathbb{R}^{m(n+1)}} \omega(z)\, dz = 1$, $\omega(z) \geqq 0$ and put ${}_h a_i^r(x, \xi, \eta)$ $\overset{\text{def}}{=} h^{-m(n+1)} \int_{\mathbb{R}^{m(n+1)}} \omega\left(\frac{\xi' - \xi}{h}, \frac{\eta' - \eta}{h}\right) \cdot a_i^r(x, \xi', \eta')\, d\xi'\, d\eta'$ and similarly for the other terms. The reader sees easily that ${}_h a_i^r, {}_h a^r, {}_h A^r$ satisfy the growth conditions in question independently of h. The conditions (3.2.8) are satisfied in the classical sense and we get the corresponding functional Φ_h. We have

$$\begin{aligned} &\frac{1}{t}[\Phi_h(u + tv) - \Phi_h(u)] \\ &= \frac{1}{t}\int_0^t d\tau \int_\Omega \left[{}_h a_i^r(x, u + \tau v, \nabla u + \tau\, \nabla v) \frac{\partial v_r}{\partial x_i} + {}_h a^r(x, u + \tau v, \nabla u + \tau\, \nabla v)\, v_r \right] dx \\ &\quad - \int_\Omega \left(f_i^r \frac{\partial v_r}{\partial x_i} - f^r v_r \right) dx + \frac{1}{t}\int_0^t d\tau \int_{\partial\Omega} {}_h A^r(x, u + \tau v)\, v_r\, dS - \int_{\partial\Omega} g^r v_r\, dS. \end{aligned} \tag{3.2.14}$$

We can make h in (3.2.14) tend to zero and, for $t \to 0$, we get the result.

The conditions of the first (and second) differential of Φ are basic in these lecture notes because of the regularity of weak solutions. What growth conditions (3.2.5), (3.2.6) are sufficient for the differentiability? It is easy to see, with regard to the imbedding theorems from Chapter 2, that if $p < n$ and $\frac{1}{q} = \frac{1}{p} - \frac{1}{n}$, then

$$(3.2.15)\qquad \left|\frac{\partial F}{\partial \eta_i^r}\right| \leqq F(x) + c|\eta|^{p-i} + c|\xi|^{\frac{np-n}{n-p}}, \quad F \in L^{p^*}(\Omega),$$

if $p = n$ then

$$(3.2.16)\qquad \left|\frac{\partial F}{\partial \eta_i^r}\right| \leqq F(x) + c|\eta|^{p-1} + c|\xi|^{\alpha},$$

$$F \in L^{p^*}(\Omega), \quad \alpha \text{ arbitrary},$$

if $p > n$, then

$$(3.2.17)\qquad \left|\frac{\partial F}{\partial \eta_i^r}\right| \leqq c(|\xi|)\,(F(x) + |\eta|^{p-1}),$$

$F \in L^{p^*}(\Omega)$, where $c(s)$ is a continuous function in $\overline{\mathbb{R}^1_+}$.

Similarly we have

$$(3.2.18)\qquad \left|\frac{\partial F}{\partial \xi_r}\right| \leqq G(x) + c|\eta|^{\frac{np+p-r}{n}} + c|\xi|,^{\frac{np+p-n}{n-p}} \qquad G \in L^{\frac{np}{np+p-n}}(\Omega),$$

for $p < n$, we have

$$(3.2.19)\qquad \left|\frac{\partial F}{\partial \xi_r}\right| \leqq G(x) + c|\xi|^{\alpha} + c|\eta|^{\beta},$$

for $p = n$, where α is arbitrary, $\beta < p$, $G \in L^{\gamma}(\Omega)$, $\gamma > 1$ and, finally, it is

$$(3.2.20)\qquad \left|\frac{\partial F}{\partial \xi_r}\right| \leqq (|\xi|)\,(G(x) + c\,|\eta|^{p}), \quad G \in L^1(\Omega),$$

for $p > n$.

In the same way the reader can find such conditions for $\frac{\partial G}{\partial \xi_r}$. We shall omit such generalizations in the sequel. All such growths are called *controllable*. If, for example, $p = 2$, then in (3.2.20) the growth for $|\eta|$ is $\frac{n+2}{2} < 2$ $(n \geqq 3)$, we thus never get the equation

$$(3.2.21)\qquad -\Delta u = f + |\nabla u|^2.$$

The equation (3.2.21) is an example of an *uncontrollable* growth. We shall not consider such growths in these lecture notes. For such questions, see for example M. GIAQUINTA [36].

3.2.22. PROPOSITION. *Let the potential* (3.2.3) *exist for the problem. Then* $u \in [W^{1,p}(\Omega)]^m$ *is a weak solution to the general problem from Definition* 3.1.10 *if* $D\Phi(u, v) = 0\ \forall v \in V$.

Such a point u, at which $D\Phi(u, v) = 0$, $\forall v \in V$, is called the *critical point*. From the abstract point of view the reader can consult other literature; as far as the calculus of variations is concerned, see [29], [5], [10] ... Considering a boundary value problem we define $F(w) \overset{\text{def}}{=} \Phi(u^0 + w)$ on V and we look for $\min_V F(w)$. We use this approach in the next without repeating it.

3.2.23. PROPOSITION. *Let V be a Banach space and F a functional $V \to \mathbb{R}^1$. If F attains its minimum at u and is differentiable here, then u is a critical point.*

3.2.24. DEFINITION. A functional F defined on a Banach space V is called *coercive* if

$$\lim_{\|v\| \to \infty} F(v) = \infty . \tag{3.2.25}$$

3.2.26. DEFINITION. A functional F defined on a Banach space V is called *weakly lower semicontinuous* if $u_n \rightharpoonup u$ (weak convergence) $\Rightarrow \underline{\lim}_{n \to \infty} F(u_n) \geqq F(u)$.

3.2.27. THEOREM. *Let a functional F be coercive and weakly lower semicontinuous on a reflexive Banach space V. Then F attains its minimum on V.*

Proof. Let $c = \inf_{v \in V} F(v)$. Let v_n be a *minimizing sequence*: $\lim_{n \to \infty} F(v_n) = c$. The coerciveness implies $\|v_n\| \leqq c_1$. Since a Banach space is reflexive iff every closed ball is weakly compact, we can suppose $v_n \rightharpoonup v$. The weak lower semicontinuity implies $c = \lim_{n \to \infty} F(v_n) \geqq F(v) \geqq c$.

3.2.28. THEOREM. *Let a functional F in a Banach space V be convex and bounded. Then it is weakly lower semicontinuous.*

Proof. Let $u_n \rightharpoonup u$. There exists $a > 0$ such that $\|w - u\| < 1 \Rightarrow F(w) < a$. Put $w = u + \omega$ and $F(w) \overset{\text{def}}{=} \Phi(\omega)$. Let $\varepsilon > 0$. For $\|\omega\| < \varepsilon$, we have

$$\begin{aligned} \Phi(\omega) &= \Phi\left((1 - \varepsilon)\, O + \varepsilon \frac{\omega}{\varepsilon}\right) \leqq (1 - \varepsilon)\, \Phi(O) + \varepsilon \Phi\left(\frac{\omega}{\varepsilon}\right) \\ &\leqq \Phi(O) + \varepsilon(a - \Phi(O)). \end{aligned} \tag{3.2.29}$$

On the other hand,

$$\begin{aligned} \Phi(O) &= \Phi\left(\frac{\omega}{1 + \varepsilon} + \left(1 - \frac{1}{1 + \varepsilon}\right)\left(-\frac{\omega}{\varepsilon}\right)\right) \\ &\leqq \frac{1}{1 + \varepsilon}\, \Phi(\omega) + \left(1 - \frac{1}{1 + \varepsilon}\right) \Phi\left(-\frac{\omega}{\varepsilon}\right) \\ &\leqq \frac{1}{1 + \varepsilon}\, \Phi(\omega) + \left(1 - \frac{1}{1 + \varepsilon}\right) a, \end{aligned} \tag{3.2.30}$$

hence $(1 + \varepsilon)\, \Phi(O) \leqq \Phi(\omega) + \varepsilon a$, and thus

$$\Phi(O) - \Phi(\omega) \leqq \varepsilon(a - \Phi(O)). \tag{3.2.31}$$

Therefore F is continuous in u (and everywhere). Suppose that $\lim_{n \to \infty} F(u_n) < F(u)$. Then $\lim_{n \to \infty} F(u_n) \leqq A < F(u)$. But the set $M = \{v \mid F(v) \leqq A\}$ is convex and, by the continuity of F, closed, hence weakly closed, which contradicts the assumption $\lim_{n \to \infty} F(u_n) < F(u)$.

3.2.32. THEOREM. *Let the functional F on V be differentiable and let*

$$F(u+h) - F(u) \geqq DF(u, h) \tag{3.2.33}$$

or

$$DF(u+h, h) - DF(u, h) \geqq 0 \tag{3.2.34}$$

(in this case let $DF(u+th, h)$ be continuous). Then F is weakly lower semi-continuous.

Proof. Let $u + h_n \rightharpoonup u \Rightarrow \lim_{n\to\infty} F(u+h_n) \geqq F(u) + \lim_{n\to\infty} DF(u, h_n) = F(u)$.
Or $F(u+h_n) - F(u) = \int_0^1 DF(u+th_n, h_n)\,dt = \int_0^1 [DF(u+th_n, h_n) - DF(u, h_n)]\,dt + DF(u, h_n) \geqq DF(u, h_n)$ and we come to (3.2.33).

3.2.35. THEOREM. *Let there exist the second differential. For F on V $\forall u \in V$ and let $D^2F(u, v, v) \geqq 0$. Suppose that $D^2F(u+th, h, h)$ is a continuous function. Then F is convex.*

Proof. $DF(u+h, h) - DF(u, h) = \int_0^1 D^2F(u+th, h, h)\,dt \geqq 0$.

3.2.36. THEOREM. *Let F on V be differentiable and such that $DF(u+h, h) - DF(u, h) > 0$ for $h \neq 0$. Then there exists at most one critical point.*

Proof is clear.

3.2.37. THEOREM. *Let Φ be a twice differentiable functional on a Hilbert space H. Let $D^2\Phi(u, h, k)$ be a continuous map from H to $\mathbb{R}^1$. Then $D^2\Phi(u, h, k) = D^2\Phi(u, k, h)$. Let*

$$m\|h\|^2 \leqq D^2\Phi(u, h, h). \tag{3.2.38}$$

Then Φ is coercive and weakly lower semicontinuous.

Proof. Looking for $\Phi(u + th + \tau k)$, then the symmetry of $D^2\Phi(u, h, k)$ follows from the calculus. We have $\Phi(u) - \Phi(u_0) = \int_0^1 D\Phi(u_0 + t(u-u_0), u-u_0)\,dt$
$= \int_0^1 [D\Phi(u_0 + t(u-u_0), u-u_0) - D\Phi(u_0, u-u_0)]\,dt + D\Phi(u_0, u-u_0)$
$= \int_0^1 t\,dt \int_0^1 D^2\Phi(u_0 + t\tau(u-u_0), u-u_0, u-u_0)\,d\tau + D\Phi(u_0, u-u_0)$
$\geqq \frac{m}{2}\|u-u_0\|^2 - c_1\|u-u_0\|$. So we got coerciveness. As we have proved also (3.2.33), we have the desired result.

3.2.39. THEOREM. *Let the functions $F(x, \xi, \eta)$ and $G(x, \xi)$, that fulfil the conditions* (3.2.1) *and* (3.2.2), *be convex in $\mathbb{R}^{m(n+1)}$ and $\mathbb{R}^m$, respectively. Then the functional* (3.2.3) *is convex. Or, let the conditions* (3.2.5), (3.2.6) *be satisfied* *and let*

(3.2.40)
$$\left[\frac{\partial F}{\partial \xi_i^r}(x, \xi', \eta') - \frac{\partial F}{\partial \xi_i^r}(x, \xi, \eta)\right](\eta_i^{r\prime} - \eta_i^r) + \left[\frac{\partial F}{\partial \xi_r}(x, \xi', \eta') - \frac{\partial F}{\partial \xi_r}(x, \xi, \eta)\right](\xi_y' - \xi_r) \geqq 0,$$

(3.2.41)
$$\left[\frac{\partial G}{\partial \xi_r}(x, \xi') - \frac{\partial G}{\partial \xi_r}(x, \xi)\right](\xi_r' - \xi_r) \geqq 0.$$

Then the functional (3.2.3) *is convex.*

3.2.42. THEOREM. *Let $p \geqq 2$, let there exist second continuous derivatives of F, G with respect to η, ξ, and let*

(3.2.43)
$$\sum_{r,i,s,j}\left|\frac{\partial^2 F}{\partial \eta_i^r \partial \eta_j^s}\right| + \sum_{r,i,s}\left|\frac{\partial^2 F}{\partial \eta_i^r \partial \xi_s}\right| + \sum_{r,s}\left|\frac{\partial^2 F}{\partial \xi_r \partial \xi_s}\right| \leqq c + c|\xi|^{p-2} + c|\eta|^{p-2},$$

(3.2.44)
$$\frac{\partial^2 F}{\partial \eta_i^r \partial \eta_j^s}\tilde{\eta}_i^r \tilde{\eta}_j^s + 2\frac{\partial^2 F}{\partial \xi_s \partial \eta_i^r}\tilde{\eta}_i^r \tilde{\xi}_s + \frac{\partial^2 F}{\partial \xi_r \partial \xi_s}\tilde{\xi}_r \tilde{\xi}_s \geqq 0,$$

(3.2.45)
$$\sum_{r,s}\left|\frac{\partial^2 G}{\partial \xi_r \partial \xi_s}\right| \leqq c|\xi|^{p-2} + c,$$

(3.2.46)
$$\frac{\partial^2 G}{\partial \xi_r \partial \xi_s}\tilde{\xi}_r \tilde{\xi}_s \geqq 0.$$

Then the functional (3.2.3) *is convex.*

3.2.47. THEOREM. *Let us assume* (3.2.43), (3.2.45) *with $p = 2$, let*

(3.2.48)
$$\frac{\partial^2 F}{\partial \eta_i^r \partial \eta_j^s}\tilde{\eta}_i^r \tilde{\eta}_j^s + 2\frac{\partial^2 F}{\partial \eta_i^r \partial \xi_s}\tilde{\eta}_i^r \tilde{\xi}_s + \frac{\partial^2 F}{\partial \xi_r \partial \xi_s}\tilde{\xi}_r \tilde{\xi}_s \geqq c_1|\tilde{\eta}|^2 + \alpha(x)|\tilde{\xi}|^2, \quad c_1 > 0,\ \alpha(x) \geqq 0,\ \alpha \in L^\infty(\Omega),$$

(3.2.49)
$$\frac{\partial^2 G}{\partial \xi_r \partial \xi_s}\tilde{\xi}_r \tilde{\xi}_s \geqq \beta(x)|\tilde{\xi}|^2, \quad \beta \in L^\infty(\partial\Omega),\ \beta(x) \geqq 0,$$

and, if $V \neq [W_0^{1,2}(\Omega)]^m$, let, for $v \in V$, $\int_\Omega |\nabla v|^2\,dx + \int_\Omega \alpha(x)|v|^2\,dx + \int_{\partial\Omega} \beta(x)|v|^2\,dS = 0 \Rightarrow v = 0$. Then (3.2.38) *is valid. If $F = F(x, \eta)$, $G \equiv 0$, and $V = [W^{1,2}(\Omega)]^m$, then* (3.2.38) *is valid for $[W^{1,2}(\Omega)]^m/\mathbb{R}^m$.*

(*The quotient space with respect to constant functions.*)

The proof is based on the equivalence of the norm $\int_\Omega |\nabla v|^2\,dx + \int_\Omega \alpha(x)|v|^2\,dx + \int_{\partial\Omega} \beta(x)|v|^2\,dS$ in V and on the fact that $\int_\Omega |\nabla v|^2\,dx$ is the norm in $[W^{1,2}(\Omega)]^m/\mathbb{R}^m$. See the proof of 2.5.1.

3.2.50. THEOREM. *Let $F(x, \xi, \eta) \geqq c_1|\eta|^p - F(x)$, $G(x, \xi) \geqq -G(x)$, $F \in L^1(\Omega)$, $G \in L^1(\partial\Omega)$. Then for $p > 1$, the functional* (3.2.3) *is coercive.*

All those theorems are more or less trivial. A little bit more interesting theorem is the following one.

3.2.51. THEOREM. *Let the conditions* (3.2.5), (3.2.5) *be satisfied. Let* $\frac{\partial^2 F}{\partial \eta_i^r \partial \eta_j^s} \tilde{\eta}_i^r \tilde{\eta}_i^s \geqq 0$. *Then the functional* (3.2.3) *is weakly lower semicontinuous on* $[W^{1,p}(\Omega)]^m$, $p > 1$.

Proof. Let $h_n \rightharpoonup 0$. By virtue of the mean-value theorem, $\Phi(u + h_n) - \Phi(u) = D\Phi(u + \tau_n h_n, h_n)$. But

$$D\Phi(u + \tau_n h_n, h_n) - D\Phi(u, h_n) = \int_\Omega \left[\frac{\partial F}{\partial \eta_i^r}(x, u + \tau_n h_n, \nabla u + \tau_n \nabla h_n) - \frac{\partial F}{\partial \eta_i^r}(x, u + \tau_n h_n, \nabla u)\right] \frac{\partial h_n^r}{\partial x_i}\, dx + c_n,$$

where

$$c_n = \int_\Omega \left[\frac{\partial F}{\partial \eta_i^r}(x, u + \tau_n h_n, \nabla u) - \frac{\partial F}{\partial \eta_i^r}(x, u, \nabla u)\right] \frac{\partial h_n^r}{\partial x_i}\, dx$$
$$+ \int_\Omega \left[\frac{\partial F}{\partial \xi_r}(x, u + \tau_n h_n, \nabla u + \tau_n \nabla h_n) - \frac{\partial F}{\partial \xi_r}(x, u, \nabla u)\right] h_n^r\, dx$$
$$+ \int_{\partial\Omega} \left[\frac{\partial G}{\partial \xi_r}(x, u + \tau_n h_n) - \frac{\partial G}{\partial \xi_r}(x, u)\right] h_n^r\, dS$$

but in view of Theorem 2.3.13 and the complete continuity of the trace u from $W^{1,p}(\Omega)$ to $L^q(\partial\Omega)$, $\frac{1}{q} > \frac{1}{p} - \frac{1}{n-1}\,\frac{p-1}{p}$ (see [25]), we get $c_n \to 0$.

3.3. The application of the theory of monotone operators

The reader can find a much more complete explanation in the book [29], for example.

Let us put $u = u^0 + w$ as in Section 2 and define an operator $T: V \to V^*$ on V by putting

$$(T(w), v) \overset{\text{def}}{=} \int_\Omega \left(a_i^r(x, u, \nabla u)\frac{\partial v_r}{\partial x_i} + a^r(x, u, \nabla u)\, v_r\right) dx + \int_{\partial\Omega} A^r(x, u)\, v_r\, dS; \tag{3.3.1}$$

here $(,)$ denotes the duality between V^* and V. If we put

$$\int_\Omega f_i^r \frac{\partial v_r}{\partial x_i}\, dx + \int_\Omega f^r v_r\, dx + \int_{\partial\Omega} g^r v_r\, dS \overset{\text{def}}{=} (f, v), \tag{3.3.2}$$

then the solution of the nonlinear problem (3.1.10) is equivalent to the equation $T(w) = f$. In the last section, we solved this problem for the special case

$$T(w) = D\Phi(u^0 + w, \cdot).$$

3.3.3. DEFINITION. Let V be a reflexive Banach space. Let $u_n \rightharpoonup u$ and $T(u_n) \rightharpoonup f$. Let $\overline{\lim\limits_{n\to\infty}} (T(u_n), u_n) \leqq (f, u)$. Then $T(u) = f$. This property is called (M).

3.3.4. DEFINITION. The operator T from V to V^* is called *coercive* if $\lim\limits_{\|u\|\to\infty} \dfrac{(Tu, u)}{\|u\|} = \infty$.

3.3.5. DEFINITION. The operator T from V to V^* is called *demicontinuous* if $u_n \to u \Rightarrow T(u_n) \rightharpoonup T(u)$. It is *bounded*, if it maps bounded sets onto bounded sets.

3.3.6. THEOREM. *Let B be a reflexive and separable space (separability can be omitted). Let T be a mapping from B to B^* with the property (M), coercive, demicontinuous and bounded. Then $T(B) = B^*$ and T^{-1} is a multivalued bounded mapping (i.e. $\|T^{-1}(\{f \mid \|f\| \leqq r < \infty)\| \leqq \bar{r} < \infty$).*

Proof. First, the coerciveness implies the boundedness of T^{-1}. We look for approximate solutions using the Galerkin method. Let $\{w_n\}_{n=1}^{\infty}$ be a basis of linearly independent elements. Let B_n be spanned by $w_1, w_2, \ldots, w_n$. Let $f \in B^*$. We look for $u_n \in B_n$ such that

$$(T(u_n), v) = (f, v) \quad \forall v \in B_n. \tag{3.3.7}$$

Let $w_j^* \in B^*$ be chosen such that $(w_i^*, w_j) = \delta_{ij}$. If $\omega \in B_n$ and $\omega = \sum_{i=1}^{n} x_i w_i$, then, for $v \in B_n$, $(T(\omega), v)$ defines an element from B_n^* uniquely determined by $\sum_{i=1}^{n} y_i w_i^*$. We thus defined a mapping $x \mapsto y$, say Ax, from $\mathbb{R}^n$ to $\mathbb{R}^n$. The coerciveness of T implies

$$\lim_{|x|\to\infty} \frac{(Ax, x)}{|x|} = \infty. \tag{3.3.8}$$

The operator A is continuous. Put $Bx \overset{\text{def}}{=} x - \varepsilon(Ax - y)$, $\varepsilon > 0$, $y \in \mathbb{R}^n$, where $(f, \omega) = (y, x)$. The relation (3.3.8) implies the existence of $R > 0$ so large and $\varepsilon > 0$ such that for $\frac{R}{2} \leqq |x| \leqq R \Rightarrow |Bx| \leqq R$: In fact, $(Bx, Bx) = |x|^2 + \varepsilon^2|Ax - y|^2 - 2\varepsilon(x, Ax) + 2\varepsilon(x, y) \leqq |x|^2 + \varepsilon^2|Ax - y|^2 - 2\varepsilon c(|x|)|x| + 2\varepsilon c_1|x|\,|y|$, where $c(s) \to \infty$ as $s \to \infty$. Thus for R large enough and $\frac{R}{2} \leqq |x| \leqq R$, $c(|x|)\,|x| - c_1|y|\,|x| > 0$. Hence for $0 < \varepsilon \leqq \varepsilon_0$, for a suitable ε_0 and $\frac{R}{2} \leqq |x| \leqq R$ we have $\varepsilon^2|Ax - y|^2 - 2\varepsilon c(|x|)\,|x| + 2\varepsilon c_1|x|\,|y| \leqq 0$. If we choose $\varepsilon \leqq \varepsilon_0$ small enough, we get $|Bx| \leqq R$ for $|x| \leqq \frac{R}{2}$. Using Brouwer's fixed point theorem, we find that there exists a fixed point x. We thus got u_n and the coerciveness implies $\|u_n\| \leqq c < \infty$. Let $u_{n_k} \rightharpoonup u$ and $T(u_{n_k}) \rightharpoonup g$. We have $(T(u_{n_k}), u_{n_k}) = (f, u_{n_k}) \to (f, u)$ and therefore the property (M) implies $T(u) = f$.

Actually, we can weaken the conditions of Theorem 3.3.6 in different ways. This is not what we wish to do in these introductory lecture notes. Nevertheless, we mention

3.3.9. DEFINITION. A bounded operator T from B to B^*, a reflexive Banach space, is called *pseudomonotone* if $u_n \rightharpoonup u$ and

$$\overline{\lim_{n\to\infty}}\,(T(u_n), u_n - u) \leqq 0 \Rightarrow \underline{\lim_{n\to\infty}}\,(T(u_n), u_n - v) \geqq (T(u), u - v) \quad \forall v \in V.$$

3.3.10. PROPOSITION. *A pseudomonotone operator is demicontinuous.*

Proof. Let us suppose that there exist $u_n \to u$ and $T(u_n) \not\rightharpoonup T(u)$. We can suppose $T(u_n) \rightharpoonup f$. We have $\overline{\lim_{n\to\infty}}\,(T(u_n), u_n - u) = 0$, thus $\underline{\lim_{n\to\infty}}\,(T(u_n), u_n - v)$ $= (f, u - v) \geqq (T(u), u - v)$, hence $T(u) = f$, which is a contradiction.

3.3.11. THEOREM. *A pseudomonotone operator has the property (M).*

Proof. Let $u_n \rightharpoonup u$, $T(u_n) \rightharpoonup f$ and $\overline{\lim_{n\to\infty}}\,(T(u_n), u_n) \leqq (f, u)$. We have $\overline{\lim_{n\to\infty}}\,(T(u_n), u_n - u) \leqq (f, u) - (f, u) = 0$, hence $(f, u - v) \geqq \overline{\lim_{n\to\infty}}\,(T(u_n), u_n - v)$ $\geqq \underline{\lim_{n\to\infty}}\,(T(u_n), u_n - v) \geqq (T(u), u - v)$, so $T(u) = f$.

3.3.12. DEFINITION. An operator from B to B^*, is called *monotone* if $(Tu - Tv, u - v) \geqq 0 \ \forall u, v \in B$.

3.3.13. DEFINITION. An operator T from B to B^* is called *hemicontinuous*, if the mapping $\lambda \mapsto (T(u + \lambda v), v)$ is continuous.

3.3.14. THEOREM. *A hemicontinuous, monotone operator has the property (M).*

Proof. Let $u_n \rightharpoonup u$, $T(u_n) \rightharpoonup f$, and $\overline{\lim_{n\to\infty}}\,(T(u_n), u_n) \leqq (f, u)$. Thus $(f, u - v)$ $\geqq \overline{\lim_{n\to\infty}}\,(T(u_n), u_n - v) = \overline{\lim_{n\to\infty}}\,(T(u_n) - T(v), u_n - v) + (T(v), u - v) \geqq (T(v), u - v)$. We apply *Minty's trick*: let $v = u - \lambda w$, $\lambda > 0$. Then $(T(u - \lambda w), w) \leqq (f, w)$, hence $(T(u), w) \leqq (f, w)$ as $\lambda \to 0$, so $T(u) = f$.

3.3.15. EXERCISE. (i) Let $(T(u), h) = D\Phi(u, h)$ and let T be coercive so that $(Tu, u) \geqq c(\|u\|)\,\|u\|$, where $c(s) \to \infty$ as $s \to \infty$. Then $\Phi(u) \geqq c_1(\|u\|)\,\|u\|$, where $c_1(s) \to \infty$ as $s \to \infty$.

(ii) If T is strictly monotone: $(Tu - Tv, u - v) > 0$ for $u \neq v$, then T^{-1} is single-valued.

(iii) Let T be monotone and hemicontinuous. Then $Tu = f \Leftrightarrow (T(v) - f, v - u)$ $\geqq 0$, $\forall v \in V$.

(iv) Let T be monotone and hemicontinuous. Then $T^{-1}f$ is a closed, convex set.

(v) Let T be hemicontinuous, monotone from a reflexive B to B^*. Then it is pseudomonotone.

In applications, the following property is very often satisfied:

3.3.16. DEFINITION. Let T be an operator from B to B^*. It has the property (S_+) if $u_n \rightharpoonup u$ and $\overline{\lim\limits_{n\to\infty}}\,(T(u_n) - T(u), u_n - u) \leqq 0 \Rightarrow u_n \to u$.

3.3.17. THEOREM. *A bounded, demicontinuous operator with the property (S_+) from a reflexive B to B^* is pseudomonotone.*

Proof. If $u_n \rightharpoonup u$ and $\overline{\lim\limits_{n\to\infty}}\,(T(u_n), u_n - u) \leqq 0$ then $u_n \to u$, hence $\underline{\lim\limits_{n\to\infty}}\,(T(u_n), u_n - v) = (T(u), u - v)$.

3.3.18. DEFINITION. Let T be an operator from B to B^*. It has the property (S) if $u_n \rightharpoonup u$, $(Tu_n - Tu, u_n - u) \to 0 \Rightarrow u_n \to u$.

3.3.19. THEOREM. *Let B be a reflexive and separable Banach space. Let T be a bounded, demicontinuous operator from B to B^* with the property (S). Let T be coercive. Then $T(B) = B^*$ and T^{-1} is bounded.*

We leave the proof to the reader: the first step is done by the Galerkin method as in the proof of Theorem 3.3.6. Then if $u_n \rightharpoonup u$, choose $v_n \in B_n$ such that $v_n \to u$. Hence $\lim\limits_{n\to\infty}(T(u_n), u_n - u) = \lim\limits_{n\to\infty}(T(u_n), u_n - v_n) = (f, u_n - v_n) = 0$, and thus $u_n \to u$.

In the end of this rather abstract formalism we reprove a special case of the general Theorem 3.3.6 by a method that is also an approximate method.

3.3.20. DEFINITION. An operator T from a Hilbert space V to itself is called *strongly monotone* if

$$(T(u) - T(v), u - v) \geqq \lambda \|u - v\|^2 \tag{3.3.21}$$

and Lipschitz continuous if

$$\|T(u) - T(v)\| \leqq \Lambda \|u - v\|. \tag{3.3.22}$$

3.3.23. THEOREM. *Let the operator T be strongly monotone and Lipschitz continuous. Then $T(V) = V$, T^{-1} is also Lipschitz continuous, and*

$$\|T^{-1}f - T^{-1}g\| \leqq \frac{1}{\lambda}\|f - g\|. \tag{3.3.24}$$

Let $0 < \varepsilon < \dfrac{2\lambda}{\Lambda^2}$ *and define*

$$Au \overset{\text{def}}{=} u - \varepsilon(Tu - f). \tag{3.3.25}$$

The operator A is contractive with $\alpha = (1 - 2\varepsilon\lambda + \varepsilon^2 \Lambda^2)^{1/2}$ (*the minimum of* α *is* $\left(1 - \frac{\lambda^2}{\Lambda^2}\right)^{1/2}$).

The main step in the proof: $\|Au - Av\|^2 = \|u - v\|^2 + \varepsilon^2 \|Tu - Tv\|^2 - 2\varepsilon(u - v, Tu - Tv) \leqq \|u - v\|^2 (1 + \varepsilon^2 \Lambda^2 - 2\varepsilon\lambda)$.

Let us now turn to elliptic systems. We begin with the easiest case of the quadratic growth of the coefficients a_i^r, a^r, and A^r. Let us suppose that a_i^r and a^r are once continuously differentiable with respect to ξ, η for almost all x in Ω and that

$$\sum_{r,i,s,j} \left|\frac{\partial a_i^r}{\partial \eta_j^s}\right| + \sum_{r,i,s} \left|\frac{\partial a_i^r}{\partial \xi_s}\right| + \sum_{r,s,j} \left|\frac{\partial a_r}{\partial \eta_j^s}\right| + \sum_{r,s} \left|\frac{\partial a_r}{\partial \xi_s}\right| \leqq c < \infty. \tag{3.3.26}$$

Let further

$$\frac{\partial a_i^r}{\partial \eta_j^s}(x, \xi, \eta)\, \zeta_i^r \zeta_j^s \geqq c|\zeta|^2, \quad c > 0, \tag{3.3.27}$$

$$a_i^r(x, \xi, \eta)\, \eta_i^r + a^r(x, \xi, \eta)\, \xi_r \geqq c_1|\eta|^2 + c_2(x)\, |\xi|^2 - c_3, \quad c_1 > 0, \tag{3.3.28}$$

where $c_2(x) \geqq 0$, $c_2 \in L^\infty(\Omega)$.

Let us assume that the function $A^r(x, \xi)$ has a continuous derivative with respect to ξ almost everywhere on $\partial\Omega$ and that

$$\sum_{r,s} \left|\frac{\partial A^r}{\partial \xi_s}\right| \leqq c, \tag{3.3.29}$$

and

$$A^r(x, \xi)\, \xi_r \geqq c_4(x)\, |\xi|^2 - c_5, \tag{3.3.30}$$

$c_4(x) \geqq 0$, $c_4 \in L^\infty(\partial\Omega)$.

3.3.31. THEOREM. *Let us consider a weak solution according to Definition 3.1.10. If* $V \neq [W_0^{1,2}(\Omega)]^m$ *then let us suppose that* $v \in V$, $\int_\Omega (|\nabla v|^2 + c_2(x)\, |v|^2)\, dx + \int_{\partial\Omega} c_4(x)\, |v|^2\, dS = 0 \Rightarrow v = 0$. *Then the operator T defined in* (3.3.1) *is coercive, bounded, continuous and it satisfies the condition* (*S*). *If further the condition*

$$\frac{\partial a_i^r}{\partial \eta_j^s}(x, \xi, \eta)\, \tilde{\eta}_i^r \tilde{\eta}_j^s + \frac{\partial a_i^r}{\partial \xi_s}(x, \xi, \eta)\, \tilde{\eta}_i^r \tilde{\xi}_s + \frac{\partial a^r}{\partial \eta_j^s}(x, \xi, \eta) \tilde{\xi}_r \tilde{\eta}_j^s + \frac{\partial a^r}{\partial \xi_s}(x, \xi, \eta)\, \tilde{\xi}_r \tilde{\xi}_s > 0 \tag{3.3.32}$$

holds for $(\tilde{\xi}, \tilde{\eta}) \neq 0$, *then the solution is unique.*

Proof. The continuity and boundedness follow from the properties of the Nemyckiĭ operator. The proof of the coerciveness is the same as in the proof of Theorem 3.2.47.

Proof of the condition (*S*): Let $u^n = u^0 + w^n \rightharpoonup u$. We have

$$(3.3.33)\qquad (T(w^n) - T(w), w^n - w)$$

$$= \int_\Omega \mathrm{d}x \int_0^1 \frac{\partial a_i^r}{\partial \eta_j^s}(x, u + t(u^n - u), \nabla u + t(\nabla u^n - \nabla u))\,\mathrm{d}t$$

$$\times \left(\frac{\partial u_r^n}{\partial x_i} - \frac{\partial u_r}{\partial x_i}\right)\left(\frac{\partial u_s^n}{\partial x_j} - \frac{\partial u_s}{\partial x_j}\right) + c_n \geqq c \int_\Omega |\nabla(u^n - u)|^2 \mathrm{d}x + c_n$$

where

$$c_n = \int_\Omega \mathrm{d}x \int_0^1 \frac{\partial a_i^r}{\partial \xi_s}\,\mathrm{d}t \left(\frac{\partial u_r^n}{\partial x_i} - \frac{\partial u_r}{\partial x_i}\right)(u_s^n - u_s)$$

$$+ \int_\Omega (a^r(x, u^n, \nabla u^n) - a^r(x, u, \nabla u))\,(u_r^n - u_r)\,\mathrm{d}x$$

$$+ \int_{\partial\Omega} [A^r(x, u^n) - A^r(x, u)]\,(u_r^n - u_r)\,\mathrm{d}S.$$

Using the compactness of the imbedding as in the proof of Theorem 3.2.47, we get $c_n \to 0$ and thus $u_n \to u$ in $[W^{1,2}(\Omega)]^m$ because of (3.3.33).

3.3.34. REMARK. For the Dirichlet problem (and in the analogous way for the other problems) we can replace the condition (3.3.28) by

$$(3.3.35)\qquad a_i^r(x, \xi, \eta)\,\eta_i^r + a^r(x, \xi, \eta)\,\xi_r \geqq c_1|\eta|^2 - c_2|\xi|^\alpha - c_3$$

with $1 < \alpha < 2$.

3.3.35. THEOREM. *Let* $a_i^r(x, \xi, \eta) = a_i^r(x, \eta)$, $a^r \equiv 0$, $A^r \equiv 0$. *Let* (3.3.27) *be satisfied. Then the operator T from* 3.3.1. *is defined on* $[W^{1,2}(\Omega)]^m/\mathbb{R}^m$, *it is bounded, continuous, coercive and it satisfies the condition* (*S*). *The necessary and sufficient condition for the existence of a unique solution from* $[W^{1,2}(\Omega)]^m/\mathbb{R}^m$ *is*

$$(3.3.36)\qquad \int_\Omega f_r\,\mathrm{d}x + \int_{\partial\Omega} g_r\,\mathrm{d}S = 0, \quad r = 1, 2, \ldots, m.$$

Proof. Obviously (3.3.36) is necessary for the existence of a solution $u \in [W^{1,2}(\Omega)]^m$; clearly, such a solution is unique apart from a constant. So we can reduce the problem to the search for $\tilde{u}$, provided (3.3.36). Finally, it follows from the Poincaré inequality that $\left(\int_\Omega |\nabla u|^2\,\mathrm{d}x\right)^{1/2}$ is an equivalent norm in $[W^{1,2}(\Omega)]^m/\mathbb{R}^m$.

Let us consider the general case $1 < p < \infty$. Let us assume (3.1.3), (3.1.4), (3.1.6), and the following *coerciveness conditions*:

$$3.3.37)\qquad a_i^r(x, \xi, \eta)\,\eta_i^r + a^r(x, \xi, \eta)\,\xi_r \geqq c_1|\eta|^p + c_3(x)\,|\xi|^p - c_3(x),$$

$c_2(x) \geqq 0$, $c_2 \in L^\infty(\Omega)$, $c_3 \in L^1(\Omega)$,

$$(3.3.38)\qquad A^r(x, \xi)\,\xi_r \geqq c_4(x)\,|\xi|^p - c_5(x),$$

$c_5 \in L^1(\partial\Omega)$, $c_4(x) \geqq 0$, $c_4 \in L^\infty(\partial\Omega)$.

Let us suppose (we leave the pure Neumann problem unnoticed—we shall turn to it later) that $v \in V$ and

$$\int_\Omega (|\nabla v|^p + c_2(x)\,|v|^p)\,dx + \int_{\partial\Omega} c_4(x)\,|v|^p\,dS = 0 \Rightarrow v = 0.$$

We can suppose either the *monotony* (*strict monotony*) conditions

(3.3.39) $$[a_i^r(x, \xi', \eta') - a_i^r(x, \xi, \eta)]\,(\eta_i^{r\prime} - \eta_i^r) + [a^r(x, \xi', \eta') - a^r(x, \xi, \eta)]\,(\xi_r' - \xi_r) \geqq 0$$

(> 0 for $(\xi', \eta') \neq (\xi, \eta)$),

(3.3.40) $$[A^r(x, \xi') - A^r(x, \xi)]\,(\xi_r' - \xi_r) \geqq 0$$

or the *pseudomonotony conditions*

(3.3.41) $$[a_i^r(x, \xi, \eta') - a_i^r(x, \xi, \eta)]\,(\eta_i^{r\prime} - \eta_i^r) > 0 \quad \text{for } \eta' \neq \eta.$$

3.3.42. THEOREM. *Let us assume* (3.1.3), (3.1.4), (3.1.6), (3.3.37), (3.3.38). *Then the operator T from Definition* 3.3.1 *is continuous, bounded, and coercive on V. The condition of monotony (strict monotony) implies that the operator T is monotone (strictly monotone). The condition of pseudomonotony implies that the operator T is pseudomonotone and satisfies the condition* (S_+) $((S_+) \Rightarrow (S)$, *of course). There exists a solution u of the problem, which is unique if the condition of strict monotony holds.*

Proof. For the coerciveness, see the foregoing proof. The only non-trivial fact is the pseudomonotony. So let $w^n \rightharpoonup w$ and put $u^n = u^0 + w^n$. Let $\overline{\lim}_{n\to\infty} (T(w^n), w^n - w) \leqq 0$. Then

(3.3.43) $$(T(w^n) - T(w), w^n - w) = \int_\Omega [a_i^r(x, u^n, \nabla u^n) - a_i^r(x, u, \nabla u)] \left(\frac{\partial w_r^n}{\partial x_i} - \frac{\partial w_r}{\partial x_i}\right) dx + c_n.$$

As above, the compactness of the imbeddings implies $c_n \to 0$ (see the proof of 3.3.31). We have

(3.3.44) $$\int_\Omega [a_i^r(x, u^n, \nabla u^n) - a_i^r(x, u, \nabla u)] \left(\frac{\partial w_r^n}{\partial x_i} - \frac{\partial w_r}{\partial x}\right) dx$$
$$= \int_\Omega [a_i^r(x, u^n, \nabla u^n) - a_i^r(x, u^n, \nabla u)] \left(\frac{\partial w_r^n}{\partial x_i} - \frac{\partial w_r}{\partial x_i}\right) dx$$
$$+ \int_\Omega [a_i^r(x, u^n, \nabla u) - a_i^r(x, u, \nabla u)] \left(\frac{\partial w_r^n}{\partial x_i} - \frac{\partial w_r}{\partial x_i}\right) dx.$$

Once more, it follows from the compactness of the imbedding $W^{1,p}(\Omega) \subsetneq L^p(\Omega)$ that the second term on the right-hand side of (3.3.44) tends to zero as $n \to \infty$. Hence for $f_n(x) \overset{\text{def}}{=} [a_i^r(x, u^n, \nabla u^n) - a_i^r(x, u, \nabla u)] \left(\frac{\partial w_r^n}{\partial x_i} - \frac{\partial w_r}{\partial x_i}\right) \geqq 0$ we have $\int_\Omega f_n(x)\,dx \to 0$. We can suppose that $f_n(x) \to 0$ and $u^n(x) \to u(x)$ in $M \subset \Omega$, where $|\Omega \setminus M| = 0$. We can also suppose that $\nabla u(x)$ is finitely defined on M. We assert that we have $\lim_{n\to\infty} \nabla u^n(x) = \nabla u(x)$ for $x \in M$. It follows from (3.3.37) that

(3.3.45)

$$\begin{aligned} f_n(x) = {} & a_i^r(x, u^n, \nabla u^n) \frac{\partial u_r^n}{\partial x_i} + a^r(x, u^n, \nabla u^n)\, u_r^n \\ & - a^r(x, u^n, \nabla u^n)\, u_r^n - a_i^r(x, u^n, \nabla u^n) \frac{\partial u_r}{\partial x_i} \\ & - a_i^r(x, u^n, \nabla u) \left(\frac{\partial u_r^n}{\partial x_i} - \frac{\partial u_r}{\partial x_i}\right) \\ \geqq {} & c_1 |\nabla u^n(x)|^p - \tilde{c}(x) - c_6 |u^n(x)|^p - c_6 |u(x)|\, |\nabla u^n(x)|^{p-1} \\ & - c_6 |u^n(x)|^{p-1}\, |\nabla u(x)| - c_6 |\nabla u^n(x)|^{p-1}\, |\nabla u(x)| \\ & - c_6 |u^n(x)|^{p-1}\, |\nabla u^n(x)| - c_6 |\nabla u(x)|^{p-1}\, |\nabla u^n(x)|, \end{aligned}$$

where $\tilde{c} \in L^1(\Omega)$ is finite in M. (We change M on a set of measure zero if necessary.) It follows from (3.3.45) that $|\nabla u^n(x)|$ is bounded; from every subsequence, we can then extract another, convergent for $x \in M$. But by virtue of (3.3.41) such a limit is $\nabla u(x)$, hence we have the assertion. But it follows also from (3.3.45), since $f_n \to 0$ in $L^1(\Omega)$, that $\forall \varepsilon > 0$, $\exists \delta > 0$ such that $|M| < \delta$ $\Rightarrow \int_M |\nabla u^n|^p \, dx < \varepsilon$ uniformly with respect to n. This, with the Jegorov theorem, gives $u^n \to u$ in $[W^{1,p}(\Omega)]^m$.

We present a method, a slight generalization of the result from [29], which enables us to consider some situations, where no growth conditions are satisfied. Let us assume $m = 1$.

3.3.46. DEFINITION. Let $a_i(x, \xi, \eta) = a_i(x, \eta)$, $|a_i(x, \eta)| \leqq a(x) + c|\eta|^{p-1}$, $1 < p < \infty$, $a \in L^{p^*}(\Omega)$, $a(x, \xi, \eta) = a(x, \xi)$, $|a(x, \xi)| \leqq c(|\xi|)\, a(x)$, let $f \in L^{p^*}(\Omega)$, $u_0 \in W^{1,p}(\Omega)$. The function $u \in W^{1,p}(\Omega) \cap L^\infty(\Omega)$ is called a *weak solution* to the *Dirichlet problem*, if

(3.3.47) $$u - u_0 \in W_0^{1,p}(\Omega),$$

(3.3.48) $$\forall v \in W_0^{1,p}(\Omega), \quad \int_\Omega \left[[a_i(x, \nabla u) \cdot \frac{\partial v}{\partial x_i} + a(x, u)\, v - fv \right] dx = 0.$$

3.3.49. DEFINITION. A function $\psi \in W^{1,p}(\Omega) \cap L^\infty(\Omega)$ is called a *subsolution* (*supersolution*) to the Dirichlet problem if

(3.3.50) $$\psi \leqq u_0 \quad (\psi \geqq u_0) \quad \text{on } \partial\Omega,$$

(3.3.51) $$\int_\Omega \left[a_i(x, \nabla \psi) \frac{\partial v}{\partial x_i} + a(x, \psi)\, v - fv \right] dx \leqq 0$$

($\geqq 0$) $\forall v \in W_0^{1,p}(\Omega)$, $v \geqq 0$ almost everywhere in Ω.

3.3.52. THEOREM. *Let us assume the condition of pseudomonotony and*

(3.3.53) $$a_i(x, \eta)\, \eta_i \geqq c_1 |\eta|^p - c(x), \quad c \in L^{p^*}(\Omega).$$

Let there exist a subsolution ϕ and a supersolution ψ to the problem. Then there exists a solution u such that $\phi \leqq u \leqq \psi$ in Ω.

Proof. Let

$$\hat{a}(x, \xi) = \begin{cases} a(x, \phi(x)) & \text{for} \quad \xi \leqq \phi(x), \\ a(x, \xi) & \text{for} \quad \phi(x) \leqq \xi \leqq \psi(x), \\ a(x, \psi(x)) & \text{for} \quad \psi(x) \leqq \xi. \end{cases}$$

Let u be a weak solution to the Dirichlet problem with $\hat{a}$ Such solution does exist according to Theorem 3.3.42. Let $v = (u - \phi)^-$. We get from (3.3.48) that

$$\int_\Omega \left[a_i(x, \nabla u)\frac{\partial v}{\partial x_i} + \hat{a}(x, \phi(x))\, v - fv\right] dx = 0. \tag{3.3.54}$$

The ϕ as a subsolution satisfies

$$\int_\Omega \left[a_i(x, \nabla\phi)\frac{\partial v}{\partial x_i} + \hat{a}(x, \phi(x))\, v - fv\right] dx \leqq 0; \tag{3.3.55}$$

thus substracting (3.3.55) from (3.3.54), we get

$$\int_\Omega [a_i(x, \nabla u) - a_i(x, \nabla\phi)]\frac{\partial v}{\partial x_i} dx \geqq 0, \tag{3.3.56}$$

which implies, in virtue of the pseudomonotony condition, $v = 0$ almost everywhere $\left(\dfrac{\partial v}{\partial x_i} = 0\text{, thus } v \text{ is constant, but } v = 0 \text{ on } \partial\Omega\right)$, hence $u \geqq \phi$ almost everywhere in Ω. In a similar way we get $u \leqq \psi$.

3.4. Weakly coercive operators, the Fredholm alternative

3.4.1. DEFINITION. Let T be a mapping from a Banach space V to its dual V^*. Such an operator is called *weakly coercive*, if

$$\lim_{\|u\| \to \infty} \|Tu\| = \infty. \tag{3.4.2}$$

In the linear case such an operator is given, for example, by $-\Delta u - \lambda u$, where λ is not an eigenvalue of the corresponding boundary value problem. Such an operator, if $\lambda > \lambda_1$, where λ is the first eigenvalue, is not coercive.

3.4.3. EXAMPLE. Let $1 < p < \infty$ and consider the Dirichlet problem

$$-\frac{\partial}{\partial x_i}\left[\left|\frac{\partial u}{\partial x_i}\right|^{p-2}\frac{\partial u}{\partial x_i}\right] - \lambda|u|^{p-2}\, u = f \quad \text{in } \Omega, \tag{3.4.4}$$

$u = 0$ on $\partial\Omega$. We can define an operator T_λ from (3.4.4) $\forall v \in W_0^{1,p}(\Omega)$ as

$$(T_\lambda u, v) \overset{\text{def}}{=} \int_\Omega \left[\sum_{i=1}^n \left|\frac{\partial u}{\partial x_i}\right|^{p-2}\frac{\partial u}{\partial x_i}\frac{\partial v}{\partial x_i} - \lambda\,|u|^{p-2}\, uv\right] dx. \tag{3.4.5}$$

Clearly, the operator is not weakly coercive, if there exists $u \neq 0$ such that $T_\lambda(u) = 0$; it is sufficient to consider the elements tu. This works, because the operator T_λ is $p-1$-homogeneous:

3.4.6. DEFINITION. An operator T from V to V^* is called *$\varkappa$-homogeneous*, $0 < \varkappa < \infty$, if $T(tu) = t^\varkappa T(u)$ for $t > 0$.

3.4.7. DEFINITION. Let T and T_0 be from V to V^*. Let T_0 be $\varkappa$-homogeneous. Then T_0 is called the *$\varkappa$-asymptote* to T if

(3.4.8) $$\lim_{\|u\| \to \infty} \frac{\|Tu - T_0 u\|}{\|u\|^\varkappa} = 0.$$

The main "abstract" theorem of this section is

3.4.9. THEOREM. *Let T be an operator from a separable, reflexive Banach space B to B^* which is demicontinuous and bounded. Let*

(3.4.10) $$T_t(x) \overset{\text{def}}{=} T(x) - tT(-x)$$

be weakly coercive, uniformly with respect to $t \in [0, 1]$, and let it have the property (S) for $\forall t \in [0, 1]$. Then $T(B) = B^*$ and T^{-1} is bounded.

For more details, see [29], [32]. It is possible to omit the condition of separability and boundedness.

Proof of Theorem 3.4.9: Let $\overline{\bigcup_{i=1}^{\infty} B_i} = B$, $\dim B_i = i$ and $f \in B^*$. There exists $R > 0$ such that $\|T_t u - (1-t)f\| \geqq \varepsilon > 0$ for $\|u\| = R$ and $t \in [0, 1]$. We again use the Galerkin method. Let us look for $u_n \in B_n$ such that

(3.4.11) $$(T(u_n) - f, v) = 0 \quad \forall v \in B_n.$$

If $g \in B^*$, put $\|g\|_n = \sup_{v \in B_n, \|v\|=1} (g, v)$. We assert: there exists n_0 such that for $n \geqq n_0$, $u \in B_n$, it is $\|u\| = R \Rightarrow \|T_t u - (1-t)f\|_n \geqq \delta > 0$, $t \in [0, 1]$. If not, then there exists $u_{n_k} \in B_{n_k}$ such that $\|u_{n_k}\| = R$, $t_{n_k} \to t$, $t_{n_k} \in [0, 1]$ and such that $u_{n_k} \rightharpoonup u$, $\|T_{t_{n_k}} u_{n_k} - (1 - t_{n_k}) f\|_{n_k} \to 0$. First, it follows that

(3.4.12) $$\|T_t u_{n_k} - (1-t)f\|_{n_k} \to 0.$$

Let $v_{n_k} \in B_{n_k}$ be such that $v_{n_k} \to u$. We get from (3.4.11) that $(T_t u_{n_k}, u_{n_k} - v_{n_k}) \to 0$, hence $(T_t u_{n_k}, u_{n_k} - u) \to 0$, hence $u_{n_k} \to u$, thus $\|u\| = R$ and $T_t u - (1-t)f = 0$, which is impossible.

So for B_n, $n \geqq n_0$, and for the the ball $B_R(O) \subset B_n$, the operator T_1 (considered from B_n to B_n^*, see the proof of 3.3.6) is odd, so according to the Borsuk theorem (for details, see [8] or [32]) $0 \neq \deg(T_1, B_R(O), 0) = \deg(T_0, B_R(O), 0)$, hence there exists $u_n \in B_r$ such that $\|u_n\| < R$ and (3.4.11) is satisfied. We complete the argument as in the proof of Theorem 3.3.6.

3.4.13. THEOREM. *Let T_0 be a $\varkappa$-homogeneous, demicontinuous, and bounded operator, satisfying the condition (S). Then T_0 is weakly coercive iff*

(3.4.14) $$T_0 u = 0 \Rightarrow u = 0.$$

In any case it is

(3.4.15) $$\|T_0 u\| \leqq c_1 \|u\|^\varkappa$$

and if (3.4.14) *is fulfilled, then*

(3.4.16) $$c_2 \|u\|^\varkappa \leqq \|T_0 u\|\,.$$

Proof. If $\|u\| = 1$ then $\|Tu\| \leqq c_1$; from the $\varkappa$-homogeneity it follows $T(O) = 0$ and (3.4.15). Clearly, the condition (3.4.14) is necessary for the weak coerciveness. Let as prove its sufficiency by contradiction, proving (3.4.16). So let u_n be such that $\frac{1}{n} \|u_n\|^\varkappa > \|T_0 u_n\|$. Put $v_n = \frac{u_n}{\|u_n\|}$; we get $\|T_0 v_n\| < \frac{1}{n}$. Let $v_{n_k} \rightharpoonup v$. As $(T_0 v_{n_k}, v_{n_k} - v) \to 0$, the condition (S) implies $v_{n_k} \to v$, thus $\|v\| = 1$ and $T_0(v) = 0$, which is impossible.

3.4.17. THEOREM. *Let T be a bounded, demicontinuous operator, satisfying the condition (S). Let T_0 be its $\varkappa$-asymptote. Then T_0 is bounded, continuous at* 0, $T_0(O) = 0$, *and demicontinuous.*

Proof. We have $\|T(u) - T_0(u)\| \leqq c(\|u\|)\,\|u\|^\varkappa$, where $c(s) \to 0$ as $s \to \infty$. Since $T_0(u) = t^\varkappa T_0(u)$; we have $T(O) = 0$. The homogeneity of T_0 also implies

(3.4.18) $$\|T_0(u)\| \leqq c_1 \|u\|^\varkappa,$$

thus T_0 is bounded. (3.4.18) gives the continuity of T_0 at 0. Let $u_n \to u \neq 0$ and let $\varepsilon > 0$, $v \in B$, $\|v\| \leqq 1$. We have

(3.4.19) $$\|T_0(u_n) - t^{-\varkappa} T(tu_n)\| \leqq c(t\|u_n\|)\,\|u_n\|^\varkappa,$$

(3.4.20) $$\|T_0(u) - t^{-\varkappa} T(tu)\| \leqq c(t\|u\|)\,\|u\|^\varkappa,$$

we can thus find t so large, for n large enough, that $c(t\|u_n\|)\,\|u_n\|^\varkappa < \frac{\varepsilon}{3}$, $c(t\|u\|)\,\|u\|^\varkappa < \frac{\varepsilon}{3}$, hence

$$|(T_0(u_n) - T_0(u), v)| \leqq \frac{2\varepsilon}{3} + |(t^{-\varkappa} T(tu_n) - t^{-\varkappa} T(tu), v)| < \varepsilon \quad \text{for } n$$

large enough.

3.4.21. THEOREM. *Let T be an operator from a reflexive, separable B to B^* and let it be bounded and demicontinuous. Let the operator T_t from* (3.4.10) *have the property (S) $\forall t \in [0, 1]$. Let T_0 be a $\varkappa$-asymptote to T, which is odd and satisfies the property (S). Then the necessary and sufficient condition for the validity of*

(3.4.22) $$\|u\|^\varkappa \leqq c(1 + \|Tu\|)$$

is the condition (3.4.14). *If* (3.4.22) *is satisfied, then* $T(B) = B^*$.

3.4.22. REMARK. Theorem 3.4.21 is the „Fredholm alternative“: if $T_0 u = 0 \Rightarrow u = 0$ for T_0 then $T(B) = B^*$ and (3.4.22) holds. On the other hand, if (3.4.22) holds, then $T_0 u = 0 \Rightarrow u = 0$.

3.4.23. EXERCISE. 1-homogeneous operator, which is Gâteaux-differentiable at 0, is linear.

Proof of Theorem 3.4.21: Let us first suppose that the condition (3.4.22) is not satisfied. Then there exist $u_n \neq 0$ such that $\|u_n\|^\varkappa > n(1 + \|Tu_n\|)$, hence $\|u_n\| \to \infty$ and $\|u_n\|^\varkappa > n(\frac{1}{2} + \|T_0 u_n\|)$ for n large enough. For $v_n = \dfrac{u_n}{\|u_n\|}$ it is $1 > n\left(\dfrac{1}{2\|u_n\|^\varkappa} + \|T_0 v_n\|\right)$ hence $T_0 v_n \to 0$. We can suppose $v_{n_k} \rightharpoonup v$, therefore the condition (S) for T_0 implies $T_0 v = 0$ and $\|v\| = 1$.

If, on the other hand, the condition (3.4.14) is satisfied, we get (3.4.16) according to Theorem 3.4.13 and hence (3.4.22) as well.

In order to prove $T(B) = B^*$ it is enough, due to Theorem 3.4.9, to prove the weak coerciveness of T_t, uniformly with respect to $t \in [0, 1]$. But $T_{0t}(x) = (1 + t)\, T_0(x)$.

Let us now turn back to differential equations. We assume the conditions (3.1.3), (3.1.4), (3.1.6). Let us further assume the existence of functions $b_i^r(x, \xi, \eta)$, $b^r(x, \xi, \eta)$, $B^r(x, \xi)$, satisfying the Carathéodory conditions, that are $p - 1$-homogeneous in the ccuple (ξ, η) and satisfy

(3.4.24) $$b_i^r(x, \xi, \eta)| \leqq c|\xi|^{p-1} + c|\eta|^{p-1},$$

(3.4.25) $$b^r(x, \xi, \eta)| \leqq c|\xi|^{p-1} + c|\eta|^{p-1},$$

(3.4.26) $$A^r(x, \xi)| \leqq c|\xi|^{p-1}.$$

Let us further suppose

(3.4.27) $$a_i^r(x, \xi, \eta)\, \eta_i^r \geqq c_1|\eta|^p - c_2|\xi|^p - a(x), \quad a \in L^1(\Omega),$$

the condition of pseudomonotony for a_i^r. As far as the functions b_i^r, b^r, B^r are concerned, let us suppose that they are odd in couples (ξ, η), satisfy the *conditions of asymptoticity*:

(3.4.28) $$\begin{aligned} &|t^{-p+1} a_i^r(x, t\xi, t\eta) - b_i^r(x, \xi, \eta)| \\ &\leqq c(t)\,[b(x) + |\xi|^{p-1} + |\eta|^{p-1}], \quad b \in L^{p^*}(\Omega), \end{aligned}$$

$c(t) \to 0$ as $t \to \infty$ and similarly

(3.4.29) $$\begin{aligned} &|t^{-p+1} a^r(x, t\xi, t\eta) - b^r(x, \xi, \eta)| \\ &\leqq c(t)\,[b(x) + |\xi|^{p-1} + |\eta|^{p-1}], \end{aligned}$$

(3.4.30) $$\begin{aligned} &|t^{-p+1} A^r(x, t\xi) - B^r(x, \xi)| \leqq c(t)\,[B(x) + |\xi|^{p-1}], \\ &B \in L^{p^*}(\partial\Omega), \end{aligned}$$

and satisfy the condition of pseudomonotony concerning b_i^r.

3.4.31. THEOREM. *Let all the conditions mentioned above be satisfied. Then there exists a solution to the boundary value problem and it is*

(3.4.32)
$$\|u\|^{p-1}_{[W^{1,p}(\Omega)]^m} \leqq c\left[1 + \sum_{r=1}^{m} \|f^r\|_{L^{p^*}(\Omega)} + \sum_{r=1}^{m} \|g^r\|_{L^{p^*}(\partial\Omega)} + \|u^0\|_{[W^{1,p}(\Omega)]^m}\right];$$

(we put $f_i^r = 0$*) iff the solution* $w \in V$*, such that* $\forall v \in V$:

$$\int_{\Omega} \left[b_i^r(x, w, \nabla w) \frac{\partial v_r}{\partial x_i} + b^r(x, w, \nabla w) v_r \right] dx + \int_{\partial\Omega} B^r(x, w) v_r \, dS = 0, \tag{3.4.33}$$

is only trivial.

Proof. Putting $u = u^0 + w$, we have, as usual,

$$(Tw, v) \overset{\text{def}}{=} \int_{\Omega} \left[a_i^r(x, u, \nabla u) \frac{\partial v_r}{\partial x_i} + a^r(x, u, \nabla u) v_r \right] dx + \int_{\partial\Omega} A^r(x, u) v_r \, dS, \tag{3.4.34}$$

and define

$$(T_0 w, v) \overset{\text{def}}{=} \int_{\Omega} \left[b_i^r(x, u, \nabla u) \frac{\partial v_r}{\partial x_i} + b^r(x, u, \nabla u) v_r \right] dx + \int_{\partial\Omega} B^r(x, u) v_r \, dS. \tag{3.4.35}$$

Define further ${}_t a_i^r(x, \xi, \eta) = {}_t a_i^r(x, u^0(x) + \xi, \nabla u^0(x) + \eta) - {}_t a_i^r(x, u^0(x) - \xi, \nabla u^0(x) - \eta)$ and ${}_t a^r$, ${}_t A^r$ in the same manner. Actually, for $t \in [0, 1]$ the functions ${}_t a_i^r$, ${}_t a^r$, ${}_t A^r$ satisfy the conditions (3.1.3), (3.1.4), (3.1.6) uniformly with respect to t and ${}_t a_i^r$ satisfy the pseudomonotony condition for $t \in [0, 1]$. So the operator T_t satisfies the condition (S). As far as the operator T_0 is concerned, we get the condition (S) if

$$b_i^r(x, \xi, \eta) \eta_i^r \geqq c_3 |\eta|^p - c_4 |\xi|^p - h(x), \quad h \in L^1(\Omega), \tag{3.4.36}$$

(see the proof of Theorem 3.3.42), and this follows from (3.4.27) and (3.4.28) for t large enough.

It remains to prove that T_0 is a $p-1$-asymptote to T Thus for example, if we put $t = \|u\|_{1,p}$, $v = \dfrac{u}{\|u\|_{1,p}}$, we have

$$\begin{aligned} &\|u\|_{1,p}^{-p+1} \left(\int_{\Omega} |a_i^r(x, u, \nabla u) - b_i^r(x, u, \nabla u)|^{\frac{p}{p-1}} dx \right)^{\frac{p-1}{p}} \\ &= \left(\int_{\Omega} |t^{1-p} a_i^r(x, tv, t \nabla v) - b_i^r(x, v, \nabla v)|^{\frac{p}{p-1}} dx \right)^{\frac{p-1}{p}} \\ &\leqq c_1 c(t) [1 + \|v\|_{1,p}], \end{aligned} \tag{3.4.37}$$

and similarly with the other terms because $\|w\|_{1,p} \to \infty \Leftrightarrow \|u\|_{1,p} \to \infty$.

3.5. Problems with asymptotes in resonance

For illustration, let us look for a weak solution to the problem

$$\begin{aligned} &-\frac{\partial}{\partial x_i} \left[\left| \frac{\partial u}{\partial x_i} \right|^{p-2} \frac{\partial u}{\partial x_i} \right] - \lambda_1 |u|^{p-2} u + g(u) = f \quad \text{in } \Omega, \\ &u = 0 \quad \text{on } \partial\Omega, \end{aligned} \tag{3.5.1}$$

where g is a bounded function on $\mathbb{R}^1$. Clearly, the $p - 1$-asymptote is the operator (3.5.1) with $g \equiv 0$. If λ_1 is an "eigenvalue" (if there exists a non-trivial solution to (3.5.1) with $g \equiv 0$), then the results of the preceding section do not apply. In general, in a p structure, it is still an open problem what happens. We shall thus consider $p = 2$ in this section.

The basic work in this direction is the paper by E. LANDESMAN and A. C. LAZER [37] the abstract version of which is the paper by J. NEČAS [38]. The development of this program was fascinating and it can hardly be described in these lecture notes. There is a good monograph by S. FUČIK [39] and this topic is also treated in the books [5], [29] and lecture notes by E. ZEIDLER [8]. The paper by H. BRÉZIS and L. NIRENBERG [34] has a monographical character. For the sake of simplicity, we make some restrictions on the asymptotes considered in these lecture notes and we work only with weak solutions.

Let us begin with a classical result by E. LANDESMAN and A. C. LAZER (its slight generalization).

Let coefficients $a_{ij}^{rs} \in L^\infty(\Omega)$ be given such that

(3.5.2) $$a_{ij}^{rs}\zeta_i^r\zeta_j^s \geqq c|\zeta|^2, \quad c > 0, \qquad a_{ij}^{rs} = a_{ji}^{sr},$$

coefficients $a^{rs} \in L^\infty(\Omega)$ with $a^{rs} = a^{sr}$, and coefficients $A^{rs} \in L^\infty(\partial\Omega)$ with $A^{rs} = A^{sr}$. Further let functions $h^r(x, \xi)$ be given that satisfy the Carathéodory conditions,

(3.5.3) $$|h^r(x, \xi)| \leqq c(x), \quad c \in L^2(\Omega),$$

and are such that

(3.5.4) $$\lim_{\xi_r \to \pm\infty} h^r(x, \xi) = h^r_\pm(x) \in L^2(\Omega).$$

3.5.5. DEFINITION. $w \in \operatorname{Ker} T$ iff $\forall v \in V$

(3.5.6) $$\int_\Omega \left[a_{ij}^{rs}\frac{\partial w_r}{\partial x_i}\frac{\partial v_s}{\partial x_i} + a^{rs}w_r v_s\right] dx + \int_{\partial\Omega} A^{rs}w_r v_s \, dS \overset{\text{def}}{=} (Tw, v) = 0.$$

Clearly the operator T is self-adjoint (we shall identify V with V^*) and from Theorem 2.3.4 it follows that $\dim \operatorname{Ker} T < \infty$, $\operatorname{Rang} T$ is closed, and $\operatorname{Ker} T = \operatorname{Coker} T$.

3.5.7. THEOREM. *Let u^0, f^r, g^r given as usual and let $\forall w \in \operatorname{Ker} T$, $\|w\| = 1$, the condition*

(3.5.8) $$\int_\Omega f^r w_r \, dx + \int_{\partial\Omega} g^r w_r \, dS - \int_\Omega \left[a_{ij}^{rs}\frac{\partial u_r^0}{\partial x_i}\frac{\partial w_s}{\partial x_j} + a^{rs}u_r^0 w_s\right] dx - \int_{\partial\Omega} A^{rs}u_r^0 w_s \, dS < \int_\Omega h^r_+ w_r^+ \, dx + \int_\Omega h^r_- w_r^- \, dx$$

(or > 0)

be satisfied. Then there exists a solution to the problem.

(3.5.9) $$u - u^0 \in V,$$

(3.5.10) $$\forall v \in V \quad (Tu, v) + \int_\Omega h_r(x, u)\, v_r \, dx = \int_\Omega f^r v_r \, dx + \int_{\partial\Omega} g^r v_r \, dS.$$

If

$$(3.5.11) \qquad h^r_+(x) < h^r(x, \xi) < h^r_-(x)$$
$$(\text{or} \quad h^r_-(x) < h^r(x, \xi) < h^r_+(x))$$

for $r = 1, 2, \ldots, m$, *then the condition* (3.5.8) (*with* $>$) *is also necessary. If one of the signs* $<$ *in* (3.5.11) *is replaced by* $\leqq$, *then the condition* (3.5.8) *is necessary with the sign* $\leqq$ ($\geqq$).

Proof. Define $(S\omega, v) \overset{\text{def}}{=} \int_\Omega h_r(x, u + \omega)\, v_r\, dx$, $(F, v) \overset{\text{def}}{=} \int_\Omega f^r v_r\, dx + \int_{\partial\Omega} g^r v_r\, dS - (Tu_0, v)$, $A(w) = \int_\Omega h^r_+ w^+_r\, dx - \int_\Omega h^r_- w^-_r\, dx$. (All these dualities are simply scalar products in $[W^{1,2}(\Omega)]^m$.) (Let $H_1 = \text{Coker}\, T$, $H_2 = \text{Ker}\, T$, and P_i be the corresponding orthogonal projectors. Put $P_i w = w_i$. Let us look for the solution to the equation

$$(3.5.12) \qquad \varepsilon P_2 \omega^\varepsilon + T\omega^\varepsilon + S\omega^\varepsilon = F.$$

Clearly $\omega^0 + u^0$ is the solution to our problem. For $0 < |\varepsilon| \leqq \varepsilon_0$, ε_0 small enough, the operator $\varepsilon P_2 + T$ is invertible and we have

$$(3.5.13) \qquad \|(\varepsilon P_2 + T)^{-1}\| \leqq \frac{c_1}{|\varepsilon|}.$$

The operator S is completely continuous by virtue of the compact imbedding $W^{1,p}(\Omega) \subsetneq L^p(\Omega)$ and bounded by a fixed constant: $\|S\omega\| \leqq c_2$. Hence the solution to (3.5.12) is equivalent to

$$(3.5.14) \qquad \omega^\varepsilon + (\varepsilon P_2 + T)^{-1} S\omega^\varepsilon = (\omega P_2 + T)^{-1} F,$$

which can be solved by the Schauder fixed point theorem. Now

$$(3.5.15) \qquad T_1\omega^\varepsilon = P_1(F - S\omega^\varepsilon)$$

and in virtue of the inequality $\|S\omega\| \leqq c_2$ and the invertibility of T on Coker T, we have

$$(3.5.16) \qquad \|{}_1\omega^\varepsilon\| \leqq c_3.$$

We show that for $\varepsilon > 0$ also the condition $\|{}_2\omega^\varepsilon\| \leqq c_4$ holds. Let us suppose the contrary. Then there exist $\varepsilon_n \searrow 0$ such that $\|{}_2\omega^{\varepsilon_n}\| \overset{\text{def}}{=} t_n \to \infty$. We can suppose that ${}_1\omega^{\varepsilon_n} \to {}_1\omega$ in $[L^2(\Omega)]^m$ and almost everywhere in Ω. Define $\frac{1}{t_n}\, {}_2\omega^{\varepsilon_n} = {}_2w^{\varepsilon_n}$ and calculate the scalar product of (3.5.12) with ${}_2w^{\varepsilon_n}$. We can also suppose that ${}_2w^{\varepsilon_n} \to {}_2w$ in $[W^{1,2}(\Omega)]^m$ and almost everywhere in Ω. We have

$$(3.5.17) \qquad \varepsilon_n t_n = (F, {}_2w^{\varepsilon_n}) - (S\omega^{\varepsilon_n}, {}_2w^{\varepsilon_n}).$$

Now in accord with (3.5.3), (3.5.4),

$$(3.5.18) \qquad \lim_{n\to\infty} \int_\Omega h^r(x, u^0 + {}_1\omega^{\varepsilon_n} + t_n\, {}_2w^{\varepsilon_n})\, {}_2w^{\varepsilon_n}_r\, dx = \int_\Omega h^r_+ w^+_r\, dx - \int_\Omega h^r_- w^-_r\, dx.$$

Thus

(3.5.19) $$\lim_{n\to\infty} [(F, {}_2w^{\varepsilon_n}) - (S\omega^{\varepsilon_n}, {}_2w^{\varepsilon_n})] = (F, w) - A(w) < 0$$

by virtue of (3.5.8), which contradicts the inequality $\varepsilon_n t_n > 0$. If the sign $>$ is considered in (3.5.8), we take $\varepsilon < 0$. So, clearly, we can suppose $\omega^{\varepsilon_n} \rightharpoonup \omega$ and we have $S\omega^{\varepsilon_n} \to S\omega$, thus $u = u^0 + \omega$ is a solution.

If (3.5.11) is satisfied, then for $w \in \operatorname{Ker} T$, $\|w\| = 1$, we have $(F, w) - (S\omega, w) > (F, w) - A(w)$, therefore the condition (3.5.8) is satisfied.

Let us mention one of the interesting features that follow from the theorem just proved: if we have the case when the condition (3.5.8) is necessary and sufficient, then Rang $(T + S)$ is a domain $O \subset V$ and if $K = \bar{K} \subset O$ is bounded, it follows that $(T + S)^{-1}(K)$ is also bounded and that the norms of the solution can be estimated by $|\{x \in \Omega \mid w_r(x) > 0\}|$, $|\{x \in \Omega \mid w_r(x) < 0\}|$.

Let us turn, still in the language of partial differential equations, to the case $h^r_\pm(x) = 0$.

3.5.20. THEOREM. *Let the operators T and S be defined as in Theorem* 3.5.7. *Let all the conditions of the theorem be satisfied except for* (3.5.8). *Let* $h^r_\pm = 0$ *and suppose that*

(3.5.21) $$|h^r(x, \xi)\, \xi_r| \leqq k(x) \in L^2(\Omega).$$

Let

(3.5.22) $$\lim_{\xi_r \to \pm\infty} h^r(x, \xi)\, \xi_r = H^r_\pm(x) \quad (\textit{no summation}).$$

Let

(3.5.23) $$\Omega^r_0(w) \overset{\text{def}}{=} \{x \in \Omega \mid w_r(x) = 0\},$$

(3.5.24) $$\Omega^v_\pm(w) \overset{\text{def}}{=} \{x \in \Omega \mid w_r(x) \gtrless 0\}$$

for $w \in \operatorname{Ker} T$, $\|w\| = 1$.

Suppose

(3.5.25) $$|\Omega^r_0(w)| = 0, \quad r = 1, 2, \ldots, m.$$

Put

(3.5.26) $$B(w) \overset{\text{def}}{=} \sum_{r=1}^{m} \left(\int_{\Omega^r_+(w)} H^r_+ \, dx + \int_{\Omega^r_-(w)} H^r_- \, dx \right).$$

Let $(F, w) = 0$ *for* $w \in \operatorname{Ker} T$ (*notation from the proof of Theorem* 3.5.7). *Then the sufficient condition for the existence of a solution is*

(3.5.24) $$|B(w)| > 0$$

for $w \in \operatorname{Ker} T$, $\|w\| = 1$.

Proof. We use the notation from the preceding theorem and its proof. Let $\varepsilon \searrow 0$ and let $u^\varepsilon = u^0 + \omega^\varepsilon$, where ω^ε are solutions to the equation (3.5.12). We have, as before,

(3.5.25) $$\|{}_1\omega^\varepsilon\| \leqq c_1.$$

We show again that $\|_2\omega^\varepsilon\| \leqq c_2$. Let us suppose the contrary. Then there exist $\varepsilon_n \searrow 0$ such that $\|_2\omega^{\varepsilon_n}\| \overset{\text{def}}{=} t_n \to \infty$ and we suppose that ${}_1\omega^{\varepsilon_n} \to {}_1\omega$ in $[L^2(\Omega)]^m$ and almost everywhere in Ω, $\frac{1}{t_n}\,{}_2\omega^{\varepsilon_n} \to {}_2w$ in V and almost everywhere in Ω. Calculate the scalar product of (3.5.12) with $t_n\,{}_2w^{\varepsilon_n}$. We get

$$\varepsilon_n t_n^2 = -(S\omega^{\varepsilon_n}, t_n\,{}_2w^{\varepsilon_n}). \tag{3.5.26}$$

Now

$$\lim_{n\to\infty} \int_\Omega h^r(x, u^0 + {}_1\omega^{\varepsilon_n} + t_n\,{}_2w^{\varepsilon_n})\, t_n\,{}_2w_r^{\varepsilon_n}\, dx \tag{3.5.27}$$

$$= \lim_{n\to\infty} \int_\Omega h^r(x, u^0 + {}_1\omega^{\varepsilon_n} + t_n\,{}_2w^{\varepsilon_n})\,(u_r^0 + {}_1\omega_r^{\varepsilon_n} + t_n\,{}_2w_r^{\varepsilon_n})\, dx$$

$$- \lim_{n\to\infty} \int_\Omega h^r(x, u^0 + {}_1\omega^{\varepsilon_n} + t_n\,{}_2w^{\varepsilon_n})\,(\omega_r^0 + {}_1\omega_r^{\varepsilon_n})\, dx.$$

But the second term on the right-hand side of (3.5.27) is zero, as $h_\pm^r = 0$. The first term is $B(w)$. If $B(w) > 0$ we are finished, if $B(w) < 0$ we take $\varepsilon_n \nearrow 0$; the reader also sees that $w \mapsto B(w)$ for $w \in \operatorname{Ker} T$, $\|w\| = 1$, is a continuous function. We complete the proof as in the previous case.

Let us now consider the case where the conditions (3.5.3) are replaced by

$$|h^r(x, \xi)| \leqq \delta|\xi| + c_\delta(x), \quad c_\delta \in L^2(\Omega) \quad \forall \delta > 0, \tag{3.5.28}$$

and the conditions (3.5.4) by the unilateral ones (h_+ can be ∞, for example):

$$\xi_r \geqq 0 \Rightarrow c_r^+(x) \leqq h^r(x, \xi), \quad c_r^+ \in L^2(\Omega), \tag{3.5.29}$$

$$\xi_r \leqq 0 \Rightarrow c_r^-(x) \geqq h^r(x, \xi), \quad c_r^- \in L^2(\Omega) \tag{3.5.30}$$

or

$$\xi_r \geqq 0 \Rightarrow c_r^+(x) \geqq h^r(x, \xi), \tag{3.5.31}$$

$$\xi_r \leqq 0 \Rightarrow c_r^-(x) \leqq h^r(x, \xi). \tag{3.5.32}$$

For $w \in \operatorname{Ker} T$, $\|w\| = 1$, put

$$\varliminf_{\substack{t\to\infty\\ v\to w}} \int_\Omega h^r(x, tv_r)\, v_r\, dx \overset{\text{def}}{=} A(w). \tag{3.5.33}$$

3.5.34. THEOREM. *Consider the operators T and S as in Theorem* 3.5.7. *Let us assume* (3.5.28)–(3.5.30). *Then there exists a solution to the problem if*

$$(F, w) < A(w). \tag{3.5.35}$$

Proof. As before we consider the equation (3.5.12). It follows from (3.5.28) that

$$\lim_{\|\omega\|\to\infty} \frac{\|S\omega\|}{\|\omega\|} = 0, \tag{3.5.36}$$

we thus get a solution ω^ε according to Theorem 3.4.31 or directly by the Schauder fixed point theorem. In virtue of (3.5.13) we can estimate the solution by

$$\|\omega^\varepsilon\| \leqq \frac{c_2}{|\varepsilon|}(\lambda(\delta) + \delta\|\omega^\varepsilon\|). \tag{3.5.37}$$

It follows from (3.5.15) that

(3.5.38) $$\|{}_1\omega^\varepsilon\| \leqq c(\delta) + c_3\delta(\|{}_1\omega^\varepsilon\| + \|{}_2\omega^\varepsilon\|).$$

Now, we use an important trick from the work [34]: we shall prove that

(3.5.39) $$h^r(x, \xi)\,\xi_r \geqq \frac{1}{\delta}\sum_{r=1}^{m} [h_r(x, \xi)]^2 - c_4(x) - c_{5,\delta}(x)\,|\xi|,$$

for $\forall\delta > 0$, where $c_4 \in L^1(\Omega)$, $c_{5,\delta} \in L^2(\Omega)$.

In fact, take $\xi_r \geqq 0$. It follows from (3.5.29) that

(3.5.40) $$h^r(x, \xi)\,\xi_r \geqq c_r^+(x)\,\xi_r \quad \text{(no summation)},$$

hence

(3.5.41) $$\begin{aligned} h^r(x, \xi)\,\xi_r &\geqq |h^r(x, \xi)\,\xi_r - c_r^+(x)\,\xi_r| + c_r^+(x)\,\xi_r \\ &\geqq |h^r(x, \xi)\,\xi_r| - (c_r^+(x) + |c_r^+(x)|)\,\xi_r \\ &\geqq [h^r(x, \xi)]^2\frac{1}{\delta} - \frac{1}{\delta}|h^r(x, \xi)|\,c_\delta(x) - 2|c_r^+(x)|\,\xi_r \\ &\geqq [h^r(x, \xi)]^2\frac{1}{\delta} - \xi_r(c_\delta(x) + 2|c_r^+(x)|) - \frac{1}{\delta}\,c_\delta(x)]^2; \end{aligned}$$

similarly we consider the case $\xi_r \leqq 0$. The summation of (3.5.41) over r gives (3.5.39). Let us denote by $(u, v)_0$ the scalar product in $[L^2(\Omega)]^m$.

Calculating the scalar product of (3.5.12) with ω^ε (in V), we get

(3.5.42) $$\varepsilon\|{}_2\omega^\varepsilon\|^2 + (T\omega^\varepsilon, \omega^\varepsilon) + (S\omega^\varepsilon, \omega^\varepsilon) - (F, \omega^\varepsilon) = 0.$$

For $v \in V$ it is

(3.5.43) $$(Tv, v) \geqq -c_6\|Tv\|_0^2,$$

hence it follows from (3.5.28), (3.5.39) and (3.5.43) that

(3.5.44) $$\begin{aligned} &\varepsilon\|{}_2\omega^\varepsilon\|^2 - c_6\|T_1\omega^\varepsilon\|^2 \\ &\quad + \frac{1}{\delta}(h^r(x, u^0 + \omega^\varepsilon), h^r(x, u^0 + \omega^\varepsilon))_0 \leqq c_7\|\omega^\varepsilon\| + c_{7,\delta}. \end{aligned}$$

On the other hand, we obtain

(3.5.45) $$\begin{aligned} \varepsilon\|{}_2\omega^\varepsilon\|^2 &= -(h^r(x, u^0 + \omega^\varepsilon), \omega_r^\varepsilon)_0 + (F, \omega^\varepsilon) \\ &\quad - (T\omega^\varepsilon, {}_1\omega^\varepsilon) \leqq -(h^r, \omega_r^\varepsilon)_0 + (F, \omega^\varepsilon) + c_8\|T_1\omega^\varepsilon\|^2. \end{aligned}$$

By virtue of (3.5.15) it follows

(3.5.46) $$\|T_1\omega^\varepsilon\|_0^2 \leqq c_9(1 + (h^r, h^r)_0),$$

hence

(3.5.47) $$\varepsilon\|{}_2\omega^\varepsilon\|^2 \leqq -(h^r, \omega_r^\varepsilon)_0 + (F, \omega^\varepsilon) + c_{10}(1 + (h^r, h^r)_0).$$

From (3.5.44) and (3.5.47) we get

(3.5.48) $$\left(\frac{1}{\delta} - c_6c_9\right)(h^r, h^r)_0 + \varepsilon\|{}_2\omega^\varepsilon\|^2 \leqq c_{8,\delta} + c_7\|\omega^\varepsilon\|.$$

If we combine (3.5.28), (3.5.47), and (3.5.48) we finally get

$$(3.5.49)\quad \varepsilon\left(1+\frac{c_{10}}{\dfrac{1}{\delta}-c_6c_9}\right)\|{}_2\omega^\varepsilon\|^2 \leqq c_{10,\delta}+c_{11}\left(\frac{1}{\dfrac{1}{\delta}-c_6c_9}+\delta\right)\|\omega^\varepsilon\|$$
$$+(F,\omega^\varepsilon)-(h^r,u_r^0+\omega_r^\varepsilon)_0\,.$$

Let us now suppose that $\|{}_2\omega^\varepsilon\|$ is not bounded as $\varepsilon \searrow 0$. Then there exists $\varepsilon_n \searrow 0$ such that $\|{}_2\omega^{\varepsilon_n}\| \overset{\text{def}}{=} t_n \to \infty$ and such that ${}_2w^{\varepsilon_n} \overset{\text{def}}{=} \dfrac{{}_2\omega^{\varepsilon_n}}{t_n} \to w$. First it follows from (3.5.38) that $\lim\limits_{n\to\infty}\dfrac{\|{}_1\omega^{\varepsilon_n}\|}{t_n} = 0$. Dividing (3.5.48) by t_n, we get

$$(3.5.50)\quad \overline{\lim_{n\to\infty}}\,\varepsilon_n t_n\left(1+\frac{c_{10}}{\dfrac{1}{\delta}-c_6c_9}\right) \leqq \overline{\lim_{n\to\infty}}\left[\frac{c_{10,\delta}}{t_n}+c_{11}\left(\frac{1}{\dfrac{1}{\delta}-c_6c_9}+\delta\right)\right.$$
$$+\left(F,\frac{{}_1\omega^{\varepsilon_n}}{t_n}+{}_2w^{\varepsilon_n}\right)-\left(h^r\left(x,t_n\left(\frac{u_r^0}{t_n}+\frac{{}_1\omega^{\varepsilon_n}}{t_n}+{}_2w^{\varepsilon_n}\right)\right),\right.$$
$$\left.\left.\frac{u_r^0}{t_n}+\frac{{}_1\omega^{\varepsilon_n}}{t_n}+{}_2w^{\varepsilon_n}\right)\right] \leqq c_{11}\left(\frac{1}{\dfrac{1}{\delta}-c_6c_9}+\delta\right)+(F,w)-A(w)\,.$$

If we choose δ small enough, the right-hand side of (3.5.50) is negative according to (3.5.35), which is impossible. We complete the proof as in Theorem 3.5.7.

Let us suppose, similarly as in (3.5.4), that

$$(3.5.51)\quad \lim_{\xi_r\to\pm\infty} h^r(x,\xi) = h^r_\pm(x)$$

almost everywhere in Ω, where we do not suppose, in general, any integrability of $h^r_\pm$; it can be, for example, $h^r_\pm(x) \equiv \infty$. Let us recall the notation (3.5.23), (3.5.24).

3.5.52. THEOREM. *Let us preserve the hypothesis of Theorem* 3.5.34. *Let* (3.5.51) *be satisfied and* (3.5.25) *as well. Then* $A(w) \geqq \int_{\Omega^r_+(w)} f^r_+ w_r\,dx - \int_{\Omega^r_-(w)} h^r_- w_r\,dx$.

Proof. Let $t_n \to \infty$, $v^n \to w$ in V, $\|w\| = 1$, $w \in \operatorname{Ker} T$. We can suppose $v^n(x) \to w(x)$ almost everywhere in Ω. For a fixed r it now follows from (3.5.41) and from Fatou's lemma that

$$\underline{\lim_{n\to\infty}}\int_\Omega h^r(x,t_nv^n)\,v_r^n\,dx \geqq \int_\Omega \underline{\lim_{n\to\infty}}\,h^r(x,t_nv^n)\,v_r^n\,dx = \int_{\Omega^r_+(w)} h^r_+ w_r\,dx + \int_{\Omega^r_-(w)} h^r_- w_r\,dx\,.$$

Let us turn to the abstract formulation of Theorems 3.5.7, 3.5.20, and 3.5.34. According to the definition of an asymptote from the work [38] and that of the recession function from [34], we introduce, taking into account the examples (3.5.18), (3.5.33):

3.5.53. DEFINITION. Let T be a selfadjoint, continuous operator from a Hilbert space H into H. Let $T(H)$ be closed and let $0 < \dim \operatorname{Ker} T < \infty$. Let S be another operator from H to H such that

$$\lim_{\|u\| \to \infty} \frac{\|Su\|}{\|u\|} = 0. \tag{3.5.54}$$

For $w \in \operatorname{Ker} T$, $\|w\| = 1$, we put

$$A(w) = \lim_{t \to \infty,\, u \to w} (S(tu), u) \tag{3.5.55}$$

and call it an *asymptote*; in general, we put

$$A_s(w) = \underline{\lim}_{t \to \infty,\, u \to w} (S(tu), u) \tag{3.5.56}$$

and call it a *subasymptote.*

3.5.57. THEOREM. *Let T and S be operators from Definition* 3.5.53. *Let the operator S be completely continuous. Let*

$$\|Su\| \leqq c < \infty \quad \forall u \in H \tag{3.5.58}$$

and let the asymptote $A(w)$ exist. Then if

$$(f, w) < A(w) \tag{3.5.59}$$

or

$$(f, w) > A(w), \tag{3.5.60}$$

for $f \in H$, $w \in \operatorname{Ker} T$, $\|w\| = 1$, we have $f \in \operatorname{Range}(T + S)$.

Write $\operatorname{Ker} T = H_2$, $H_1 = H \dot{-} H_2$ and let P_i be the corresponding projectors. Then for $\varepsilon \neq 0$, ε small enough, the equation

$$\varepsilon P_2 u^\varepsilon + Tu^\varepsilon + Su^\varepsilon = f \tag{3.5.61}$$

is solvable $\forall f \in H$. If the condition (3.5.59) *is satisfied, then $\|u^\varepsilon\| \leqq c < \infty$ for $\varepsilon > 0$ and from every sequence u^{ε_n}, $\varepsilon_n \to 0$, a strongly convergent subsequence can be extracted. If $u^{\varepsilon_n} \to u$ for some $\varepsilon_n \to 0$, $\varepsilon_n > 0$, then $Tu + Su = f$. If* (3.5.60) *is fulfilled, then we take $\varepsilon < 0$.*

If for every $u \in H$, $w \in \operatorname{Ker} T$, $\|w\| = 1$, the inequality

$$(w, Su) < A(w) \quad (>) \tag{3.5.62}$$

is valid then the condition (3.5.59), (3.5.60) *is necessary.*

The proof is, in fact, identical with the proof of Theorem 3.5.7.

As far as the abstract version of Theorem 3.5.34 is concerned, we shall consider, for the sake of simplicity, the following case: let T be the operator from Definition 3.5.53. Let $H \subsetneq H_0$, where H_0 is another Hilbert space and let us suppose that the identity map from H to H_0 is completely continuous. Let us suppose for simplicity that

$$\|u\|_0 \leqq \|u\|. \tag{3.5.63}$$

Let N be a continuous operator from H_0 to H_0 such that

$$\lim_{\|u\|_0 \to \infty} \frac{\|Nu\|_0}{\|u\|_0} = 0. \tag{3.5.64}$$

3.5.65. THEOREM. *Let T be the operator from Definition* 3.5.53. *Let N be a continuous operator from H_0 to H_0, satisfying* (3.5.64). *Let*

$$(Tu, u) \geqq -\alpha \|Tu\|^2 \tag{3.5.65}$$

and

$$(Nu, u)_0 \geqq \beta \|Nu\|_0^2 - \gamma \|u\|_0 - \delta(\beta), \tag{3.5.66}$$

where β is an arbitrary positive number. Let $A_s(w) = \varliminf_{\substack{t \to \infty \\ u \to w}} (N(tu), u)_0$. Let $f \in H$ and look for $u \in H$ such that, $\forall v \in H$

$$(Tu, v) + (Nu, v)_0 = (f, v). \tag{3.5.67}$$

If

$$(f, w) < A_s(w), \tag{3.5.68}$$

$\forall w \in \operatorname{Ker} T$, $\|w\| = 1$, *then there exists a solution to* (3.5.67). *If $\varepsilon \neq 0$ and ε is small enough, then the problem*

$$\varepsilon(P_2 u^\varepsilon, v) + (Tu^\varepsilon, v) + (Nu^\varepsilon, v) = (f, v) \tag{3.5.69}$$

has a solution, and $\|u^\varepsilon\| \leqq c_1 < \infty$ for $\varepsilon > 0$. Every sequence contains a convergent subsequence and every convergent subsequence tends to a solution of the problem (3.5.67). *If $A(w) = \infty$, then the problem is solvable for $\forall f \in H$.*

Proof is an analogue to that of Theorem 3.5.34. Clearly $\varepsilon P_2 + T$ is invertible for $\varepsilon \neq 0$, ε small enough, and

$$\|P_1(\varepsilon P_2 + T)^{-1}\| \leqq c_1, \tag{3.5.70}$$

$$\|P_2(\varepsilon P_2 + T)^{-1}\| \leqq \frac{c_1}{|\varepsilon|}. \tag{3.5.71}$$

Put $(Su, v) \overset{\text{def}}{=} (Nu, v)_0$. Then the solvability of (3.5.69) is equivalent to the equation

$$u^\varepsilon + (\varepsilon P_2 + T)^{-1} Su^\varepsilon = (\varepsilon P_2 + T)^{-1} f \tag{3.5.72}$$

which has a solution according the Schauder fixed point theorem. We get from (3.5.70) and (3.5.64) that

$$\|P_1 u^\varepsilon\| \leqq c_2(\delta) + c_3 \delta \|u^\varepsilon\|. \tag{3.5.73}$$

Put $v = u^\varepsilon$ in (3.5.69). We get

$$\varepsilon \|P_2 u^\varepsilon\|^2 + (Tu^\varepsilon, u^\varepsilon) + (Nu^\varepsilon, u^\varepsilon)_0 - (f, u^\varepsilon) = 0. \tag{3.5.74}$$

From (3.5.65), (3.5.66), and (3.5.74) it follows

$$\varepsilon \|P_2 u^\varepsilon\|^2 - \alpha \|Tu^\varepsilon\|^2 + \beta \|Nu^\varepsilon\|_0^2 - \gamma \|u\|_0 - \delta(\beta) \leqq \|f\| \, \|u^\varepsilon\|. \tag{3.5.75}$$

Further, we have

$$\varepsilon \|P_2 u^\varepsilon\|^2 \leqq \alpha \|TP_1 u^\varepsilon\|^2 - (Nu^\varepsilon, u^\varepsilon)_0 + (f, u^\varepsilon). \tag{3.5.76}$$

Taking $v \in P_1 H$, we get from (3.5.69) that

$$\|TP_1 u^\varepsilon\| \leqq \|f\| + \|Nu^\varepsilon\|_0, \tag{3.5.77}$$

hence

(3.5.78) $$\varepsilon\|P_2u^\varepsilon\|^2 \leqq -(Nu^\varepsilon, u^\varepsilon)_0 + (f, u^\varepsilon) + 2\alpha(\|f\|^2 + \|Nu^\varepsilon\|_0^2).$$

The combination of (3.5.74), (3.5.77), and (3.5.78) implies

(3.5.79) $$\varepsilon\frac{\beta}{\beta - 2\alpha}\|P_2u^\varepsilon\|^2$$
$$\leqq (f, u^\varepsilon) - (Nu^\varepsilon, u^\varepsilon)_0 + \|u^\varepsilon\|\left(\frac{2\alpha\gamma}{\beta - 2\alpha} + \frac{2\alpha\|f\|}{\beta - 2\alpha}\right)$$
$$+ \|f\|^2\left(2\alpha + \frac{(2\alpha)^2}{\beta - 2\alpha}\right) + \delta(\beta)\frac{2\alpha}{\beta - 2\alpha}.$$

Let us now suppose that $\|P_2u^{\varepsilon_n}\| = t_n \to \infty$, $\varepsilon_n \searrow 0$. It follows from (3.5.73) that $\frac{\|P_1u^{\varepsilon_n}\|}{t_n} \to 0$. We can suppose that $\frac{1}{t_n}P_2u^{\varepsilon_n} \overset{\text{def}}{=} w_n \to w$ in Ker T. Hence

$$0 \leqq \overline{\lim_{n\to\infty}}\, \varepsilon_n\frac{\beta}{\beta - 2\alpha}t_n \leqq (f, w) - A_s(w) + \frac{2\alpha\gamma + 2\alpha\|f\|}{\beta - 2\alpha} < 0$$

for β large enough, which is impossible. We complete the proof as in the previous theorems.

3.5.80. DEFINITION. Let S be an operator from H to H and let $\|Su\| \leqq c < \infty$ for $\forall u \in H$. Let T be the operator from Definition 3.5.53. We say that S has an *asymptote* of *second order* on Ker T if

(3.5.81) $$a(w) = \lim_{t\to\infty, u\to w}(S(tu), tu) = a(w) \in R^1$$

for $\|w\| = 1$, $w \in$ Ker T.

3.5.82. THEOREM. *Let* $\lim_{t\to\infty, u\to w} S(tu) = 0$ *for* $w \in$ Ker T, $\|w\| = 1$. *Let the operator S from Definition* 3.5.80 *be completely continuous. Let* $f \in H$, $(f, w) = 0$ $\forall w \in$ Ker T. *If* $a(w) > 0$ (< 0) *then the equation* $Tu + Su = f$ *has a solution. If* u^ε *are the solutions to the equation* $\varepsilon P_2u^\varepsilon + Tu^\varepsilon + Su^\varepsilon = f$ *(which exist for* $\varepsilon \neq 0$, ε *small enough), then* $\|u^\varepsilon\| \leqq c_1 < \infty$ *for* $\varepsilon > 0$ (< 0). *Every sequence* u^{ε_n}, $\varepsilon_n \to 0$, *contains a convergent subsequence and every convergent subsequence tends to a solution.*

Proof is an analogue of that of Theorem 3.5.20 and we leave it to the reader.

Let us turn to the situation similar as in Theorem 3.5.7 where the operator S is not compact. We shall follow the paper by J. Jarušek, J. Nečas [41]. Let us begin with an example and then follow with an abstract theorem.

So let us consider coefficients satisfying the Carathéodory conditions and the growth conditions

(3.5.83) $$|h_i^r(x, \xi, \eta)| \leqq a(x), \quad a \in L^2(\Omega),$$

(3.5.84) $$|h^r(x, \xi, \eta)| \leqq a(x),$$

(3.5.85) $$|A^r(x, \xi)| \leqq A(x), \quad A \in L^2(\partial\Omega).$$

Let the operator T be given by Definition 3.5.5, u^0, f^r, g^r be given as usual and the space V as well.

So we look for $u \in [W^{1,2}(\Omega)]^m$ such that

$$u - u^0 \in V, \tag{3.5.86}$$

$$\int_\Omega \left[a_{ij}^{rs} \frac{\partial u_s}{\partial x_j} + h_i^r(x, u, \nabla u)\right] \frac{\partial v_r}{\partial x_i} \, dx + \int_\Omega [a^{rs} u_s + g^r(x, u, \nabla u) - f^r] \, v_r \, dx + \int_{\partial\Omega} [A^{rs} u_s + A^r(x, u) - g^r] \, v_r \, dS = 0 \quad \forall v \in V. \tag{3.5.87}$$

Suppose that $\operatorname{Ker} T \neq \{0\}$ and

$$[h_i^r(x, \xi, \eta') - h^r(x, \xi, \eta)] \, (\eta_i^{\prime r} - \eta_i^r) \geqq 0 \tag{3.5.88}$$

and assume conditions of asymptoticity:

$$\lim_{\eta_i{}^r \to \pm\infty} h_i^r(x, \xi, \eta) = h_\pm^{ri}(x), \tag{3.5.88$_1$}$$

$$\lim_{\xi_r \to \pm\infty} h^r(x, \xi, \eta) = h_\pm^r(x), \tag{3.5.88$_2$}$$

$$\lim_{\xi_r \to \pm\infty} A^r(x, \xi) = A_\pm^r(x). \tag{3.5.89}$$

For $w \in \operatorname{Ker} T$, $\|w\|_{1,2} = 1$, we put, as above,

$$A(w) = \int_\Omega \left[h_+^{ri} \left(\frac{\partial w_r}{\partial x_i}\right)^+ - h_-^{ri} \left(\frac{\partial w_r}{\partial x_i}\right)_- + h_+^r w_r^+ - h_-^r w_r^-\right] dx + \int_{\partial\Omega} [A_+^r w_r^+ - A_-^r w_r^-] \, dS \tag{3.5.90}$$

and

$$(F, w) = \int_\Omega f^r w_r \, dx + \int_{\partial\Omega} g^r w_r \, dS - \int_\Omega \left[a_{ij}^{rs} \frac{\partial u_s^0}{\partial x_j} \frac{\partial w_r}{\partial x_i} + a^{rs} u_s^0 w_r\right] dx - \int_{\partial\Omega} A^{rs} u_s^0 w_r \, dS. \tag{3.5.91}$$

3.5.92. THEOREM. *Let the problem* (3.5.86), (3.5.87) *be given provided* (3.5.83)–(3.5.85), (3.5.88)–(3.5.89). *Introduce* (3.5.90), (3.5.91). *Then if*

$$(F, w) < A(w) \quad (>), \tag{3.5.93}$$

the problem has a solution.

Let us prove 3.5.92 in an abstract formulation.

3.5.94. THEOREM. *Let the operators T, S be those from Definition* 3.5.53. *Let the asymptote* (3.5.55) *exist. Let the operator S be demicontinuous and let*

$$\|Su\| \leqq c < \infty \quad \forall u \in H. \tag{3.5.95}$$

Let $G = T + S$ and suppose that the operator $G_t(x) \overset{\text{def}}{=} G(x) - tG(-x)$ satisfies the condition (S) for $t \in [0, 1]$. Let the condition

$$(F, w) < A(w) \quad (>0) \tag{3.5.96}$$

be satisfied. Then there exists a solution to the problem

$$Tu + Su = F. \tag{3.5.97}$$

If $H_2 = \operatorname{Ker} T$, $H_1 = H \dot{-} H_2$, *and* P_i *are the orthogonal projectors onto* H_i, *then for* $\varepsilon \neq 0$, ε *small enough, there exists a solution to the problem*

$$\varepsilon P_2 u^\varepsilon + Tu^\varepsilon + Su^\varepsilon = F. \tag{3.5.98}$$

Then $\|u^\varepsilon\| \leqq c_1 < \infty$ *for* $\varepsilon > 0$ *and from every sequence* u^{ε_n}, $\varepsilon_n \to 0$, *it is possible to extract a convergent subsequence. Every convergent subsequence* u^{ε_n} *tends to a solution of the problem* (3.5.97).

The proof does not differ much from the proof of Theorem 3.5.7 and we shall do it more briefly, with the notations preserved.

First, the operator $\varepsilon P_2 + T + S$ is bounded, demicontinuous, and satisfies the condition (S) as well as the operators

$$\varepsilon P_2 x + Tx + Sx - t[\varepsilon P_2(-x) + T(-x) + S(-x)], \quad 0 \leqq t \leqq 1.$$

For $\varepsilon \neq 0$ it is

$$\|u\| \leqq c(\varepsilon)(1 + \|\varepsilon P_2 u + Tu + Su\|), \tag{3.5.99}$$

so, in virtue of Theorem 3.4.21, we have a solution u^ε. Now clearly, as before,

$$\|{}_1u^\varepsilon\| \leqq c_1. \tag{3.5.100}$$

Suppose that $\|P_2 u^{\varepsilon_n}\| \overset{\text{def}}{=} t_n \to \infty$ for $\varepsilon_n > 0$, $\varepsilon_n \to 0$. Calculating the scalar product of (3.5.98) with u^ε, we get, putting $v^{\varepsilon_n} = \frac{1}{t_n} u^{\varepsilon_n}$, that

$$\varepsilon_n t_n = -\frac{1}{t_n}(T_1 u^{\varepsilon_n}, {}_1u^{\varepsilon_n}) - (St_n v^{\varepsilon_n}, v^{\varepsilon_n}) + (F, v^{\varepsilon_n}). \tag{3.5.101}$$

We can suppose that ${}_2v^{\varepsilon_n} \to w$, hence $v^{\varepsilon_n} \to w$. Now

$$\lim_{n\to\infty}\left[-\frac{1}{t_n}(T_1 u^{\varepsilon_n}, {}_1u^{\varepsilon_n}) - (St_n v^{\varepsilon_n}, v^{\varepsilon_n}) + (F, v^{\varepsilon_n})\right] = -A(w) + (F, w),$$

which is a contradiction. Take u^{ε_n}. We can extract a subsequence, denoted also by u^{ε_n}, such that $u^{\varepsilon_n} \rightharpoonup u$. Now

$$Tu^{\varepsilon_n} + Su^{\varepsilon_n} = F - \varepsilon_n P u^{\varepsilon_n} \to F,$$

hence the condition (S) implies $u^{\varepsilon_n} \to u$.

The idea of the proof of Theorem 3.5.92. We put $u = u^0 + \omega$, $\omega \in V$, and define the operators T and F by (3.5.87) (put $u = u^0 + \omega$ there) and (3.5.91). Put

$$(S\omega, v) \overset{\text{def}}{=} \int_\Omega h_i^r(x, u^0 + \omega, \nabla u^0 + \nabla\omega)\frac{\partial v_r}{\partial x_i}\,\mathrm{d}x \tag{3.5.102}$$

$$+ \int_\Omega h^r(x, u^0 + \omega, \nabla u^0 + \nabla\omega)\, v_r\,\mathrm{d}x + \int_{\partial\Omega} A^r(x, u^0 + \omega)\, v_r\,\mathrm{d}S.$$

Now, by the same argument as in the proof of Theorem 3.4.31, we get that for the operator G_t, $t \in [0, 1]$, the conditions (S) is satisfied. So let us calculate

$A(w)$. Let us consider, for example,

$$\lim_{\substack{t \to \infty \\ \omega \to w}} \int_\Omega h_i^r(x, u^0 + t\omega, \nabla u^0 + t\,\nabla\omega) \frac{\partial \omega_r}{\partial x_i} \,dx.$$

Let $t_n \to \infty$, $\omega^n \to w$ in V. We can suppose that $\omega^n(x) \to w(x)$ as well as $\dfrac{\partial \omega_r^n}{\partial x_i}(x) \to \dfrac{\partial w_r}{\partial x_i}(x)$ almost everywhere in Ω. Let $\Omega_\pm^{ri}(w) = \{x \in \Omega \mid \dfrac{\partial w_r}{\partial x_i}(x) > 0\,(<0)\}$, $\Omega_0^{ri}(w) = \{x \in \Omega \mid \dfrac{\partial w_r}{\partial x_i}(x) = 0\}$. Now, on $\Omega_+^{ri}(w)$ (for example),

$$\lim_{n \to \infty} h_i^r(x, u^0 + t_n\omega^n, \nabla u^0 + t_n\,\nabla\omega^n) = h_+^{ri}(x) \quad \text{in } L^2(\Omega_+^{ri}),$$

and we thus get the result.

Chapter 4

An excursion to approximate methods

These lecture notes are not a text concerning the approximate methods in a large extent. The only point of view that we consider is the linearization and the obvious fact that the abstract methods we used to prove the existence theorems are constructive in principle. The Galerkin method, that we shall also touch, is typical; in much larger extent, this method is treated in the book [31] and in the lecture notes [8]. We employ the book [12] as our source as far as the gradient methods are concerned. We treat also the Newton's method (from the variational point of view) which requires a regular solution. We include the secant modulus method, too, which is rapidly convergent in practice. It is surprising that the continuous analogue of the Newton method, the imbedding method or, in other words, the method of differentiable homotopy, works for weak solutions. We refer to the book by H. SCHWETLICK [42].

4.1. The Galerkin-Ritz method

See also [5]. Let, as usual,

$$[W_0^{1,2}(\Omega)]^m \subsetneq V \subsetneq [W^{1,2}(\Omega)]^m, \tag{4.1.1}$$

$$u^0 \in [W^{1,2}(\Omega)]^m, \tag{4.1.2}$$

$$f \in [L^2(\Omega)]^m, \tag{4.1.3}$$

$$g \in [L^2(\partial\Omega)]^m, \tag{4.1.4}$$

and let the functional

$$\Phi(u) \overset{\text{def}}{=} \int_\Omega [F(x, u, \nabla u) - u_i f_i]\, \mathrm{d}x + \int_{\partial\Omega} [G(x, u) - u_i g_i]\, \mathrm{d}S \tag{4.1.5}$$

be defined for $u \in [W^{1,2}(\Omega)]^m$. Let us suppose, for the sake of simplicity, that F, G are twice continuously differentiable with respect to ξ, η and such that

$$\sum_{r,i,s,j} \left| \frac{\partial^2 F}{\partial \eta_i^r \partial \eta_j^s} \right| + \sum_{r,i,s} \left| \frac{\partial^2 F}{\partial \eta_i^r \partial \xi_s} \right| + \sum_{r,s} \left| \frac{\partial^2 F}{\partial \xi_r \partial \xi_s} \right| \leqq c < \infty, \tag{4.1.6}$$

$$\sum_{r,s} \left| \frac{\partial^2 G}{\partial \xi_r \partial \xi_s} \right| \leqq c < \infty. \tag{4.1.7}$$

We look for $u \in [W^{1,2}(\Omega)]^m$ such that

$$u - u^0 \in V, \tag{4.1.8}$$

$$\forall v \in V, \quad D\Phi(u, v) = 0. \tag{4.1.9}$$

We turn immediately to an abstract version of the Ritz-Galerkin method.

4.1.10. THEOREM. *Let a functional Φ from a Hilbert space H to R^1 be twice differentiable in the Gâteaux sense (the Gâteaux differential is supposed to be a linear and bounded functional). Suppose that $D^2\Phi(u, h, k)$ is a continuous function from H to R^1 with h, k fixed. Suppose further that*

$$D^2\Phi(u, h, h) \leqq \Lambda \|h\|^2, \tag{4.1.11}$$

$$D^2\Phi(u, h, h) \geqq \lambda \|h\|^2, \quad m > 0. \tag{4.1.12}$$

Let $\bigcup_{n=1}^{\infty} H_n = H$, $H_1 \subset H_2 \subset \ldots$, $\dim H_n = n$. Then there exists a unique minimizing point u_n, u in H_n, H,

$$\min_{v \in H_n} \Phi(v) = \Phi(u_n), \qquad \min_{v \in H} \Phi(v) = \Phi(v). \tag{4.1.13}$$

(u_n is a minimizing sequence:

$$\Phi(u_n) \to \Phi(u), \tag{4.1.14}$$

$$u_n \to u \quad \textit{and} \tag{4.1.15}$$

$$\Phi(u_n) - \Phi(u) \geqq \frac{\lambda}{2} \|u_n - u\|^2.) \tag{4.1.16}$$

The proof is obvious: we only mention that evidently $u_n \rightharpoonup u$. Now

$$\begin{aligned} &\Phi(u_n) - \Phi(u) - D\Phi(u, u_n - u) \\ &= \int_0^1 \int_0^1 D^2\Phi(u + t\tau(u_n - u), u_n - u, u_n - u)\, t \, dt \, d\tau \\ &\geqq \frac{\lambda}{2} \|u_n - u\|^2. \end{aligned} \tag{4.1.17}$$

Since $D\Phi(u, v) = 0$, we have (4.1.16).

4.2. Method of steepest descent

Define the gradient of Φ in u as

$$D\Phi(u, h) \overset{\text{def}}{=} (G(u), h). \tag{4.2.1}$$

4.2.2. THEOREM. *Suppose that the conditions of Theorem* 4.1.10 *are satisfied. Let* u_n *be contructed. Put*

(4.2.3) $$u_{n+1} = u_n - \frac{1}{\Lambda} G(u_n)$$

or let $\varrho_n \geqq 0$ *be chosen in such a way that* $\Phi(u_n - \varrho_n G(u_n))$ *is minimal; put*

(4.2.4) $$u_{n+1} = u_n - \varrho_n G(u_n).$$

If u *is the critical point of the functional* Φ, *then* $u_n \to u$ *and*

(4.2.5) $$\lambda^2 \|u_n - u\|^2 \leqq 2\Lambda(\Phi(u_n) - \Phi(u)).$$

Proof. We have

(4.2.6) $$\begin{aligned}\Phi(u_{n+1}) &= \Phi(u_n) + (G(u_n), u_n - u) \\ &\quad + \tfrac{1}{2} D^2\Phi(u_n + \theta(u_{n+1} - u_n), u_{n+1} - u_n, u_{n+1} - u_n), \\ &\quad 0 < \theta < 1.\end{aligned}$$

If $u_{n+1} = u_n - \varrho G(u_n)$, then

(4.2.7) $$\Phi(u_{n+1}) \leqq \Phi(u_n) - \varrho \|G(u_n)\|^2 + \tfrac{1}{2}\varrho^2 \Lambda \|G(u_n)\|^2.$$

The quadratic polynomial on the right-hand side of (4.2.7) attains its minimum for $\varrho = \dfrac{1}{\Lambda}$, thus in both the cases of (4.2.3) and (4.2.4) we have

(4.2.8) $$\Phi(u_{n+1}) \leqq \Phi(u_n) - \frac{1}{2\Lambda} \|G(u_n)\|^2,$$

hence

(4.2.9) $$\|G(u_n)\|^2 \leqq 2\Lambda(\Phi(u_n) - \Phi(u_{n+1})).$$

But $(G(u_n), u_n - u) = (G(u_n), u_n - u) - (G(u), u_n - u) \geqq m\|u_n - u\|^2$ and therefore

(4.2.10) $$\lambda \|u_n - u\| \leqq \|G(u_n)\|.$$

4.2.11. REMARK. Use the iteration (4.2.3) in all steps. Let $H(u)$ be a linear bounded mapping from H to H defined by $(H(u)\,k, h) \overset{\text{def}}{=} D^2\Phi(u, h, k)$. Suppose that $u \to H(u)\,v$ is a continuous map from segments to H. Then

(4.2.12) $$\|G(u_{n+1})\| \leqq \left(1 - \frac{\lambda}{\Lambda}\right) \|G(u_n)\|.$$

Proof. In fact, we have

(4.2.13) $$\begin{aligned}(G(u_{n+1}), h) &= \left(\int_0^1 H(u_n + t(u_{n+1} - u_n))\,(u_{n+1} - u_n)\,\mathrm{d}t, h\right) \\ &\quad + (G(u_n), h),\end{aligned}$$

for $h \in H$, thus

(4.2.14) $$\begin{aligned}G(u_{n+1}) &= \int_0^1 H(u_n + t(u_{n+1} - u_n))\,(u_{n+1} - u_n)\,\mathrm{d}t + G(u_n) \\ &= G(u_n) - \frac{1}{\Lambda} \int_0^1 H(u_n + t(u_{n+1} - u_n))\,G(u_n)\,\mathrm{d}t \\ &\overset{\text{def}}{=} AG(u_n).\end{aligned}$$

A is a linear bounded self-adjoint operator, hence

(4.2.15) $$\|A\| \leqq 1 - \frac{\lambda}{\Lambda}.$$

4.3. The Newton method

We refer to the book by L. V. KANTOROVIČ and G. P. AKILOV [43]. We define the Hessian $H(u)$ as above by

(4.3.1) $$D^2\Phi(u, h, k) \overset{\text{def}}{=} (H(u)\, k, h).$$

Let us mention the obvious fact that the bilinear form $D^2\Phi(u, h, k)$ is symmetric:

(4.3.2) $$D^2\Phi(u, h, k) = D^2\Phi(u, k, h),$$

hence, in virtue of (4.1.11), (4.1.12), we suppose all the time that it is bounded. So $H(u)$ is a linear bounded self-adjoint positive definite operator. Let u_n be the n-th iterate approximating the solution to

(4.3.3) $$G(u) = 0.$$

We have $0 = G(u) = G(u_n) + \int_0^1 H(u_n + t(u_n - u))\,(u_n - u)\,dt =$ (provided that $H(u)\,h$ is continuous as the map $u \to H(u)\,h) \doteq G(u_n) + H(u_n)\,(u - u_n)$, thus the natural (Newton) iteration is

(4.3.4) $$u_{n+1} = u_n - H^{-1}(u_n)\, G(u_n).$$

This iteration is defined also in the case without potential.

Let us consider, in general,

(4.3.5) $$u_{n+1} = u_n - \varrho H^{-1}(u_n)\, G(u_n).$$

For the sake of simplicity and for a unique point of view, we confine ourselves to the potential case.

4.3.6. THEOREM. (*The Newton method with a short step.*) *Let the hypothesis of Theorem* 4.2.2 *be satisfied. Let* $\varrho \in \left(0, \frac{2\lambda}{\Lambda}\right)$. *Then*

(4.3.7) $$\lambda^2\|u_n - u\|^2 \leqq \frac{\Lambda^2}{\varrho\left(\lambda - \frac{\varrho\Lambda}{2}\right)} (\Phi(u_n) - \Phi(u_{n+1})).$$

Thus $u_n \to u$, *where* $G(u) = 0$. *The best constant in* (4.3.7) *is* $\varrho = \frac{\lambda}{\Lambda}$ *and it gives* $m^2\|u_n - u\|^2 \leqq \frac{2\Lambda^3}{\lambda^2}(\Phi(u_n) - \Phi(u_{n+1}))$.

Proof. We have $\|H^{-1}(v)\| \leqq \frac{1}{\lambda}$ and put $H^{-1}G(u_n) \overset{\text{def}}{=} \omega_n$. We have, as above,

(4.3.8) $$\begin{aligned}\Phi(u_{n+1}) &= \Phi(u_n) - \varrho(G(u_n), \omega_n) + \tfrac{1}{2}\varrho^2(H(u_n - \theta\varrho\omega_n)\,\omega_n, \omega_n)\\ &\leqq \Phi(u_n) - \varrho\lambda\|\omega_n\|^2 + \tfrac{1}{2}\varrho^2\Lambda\|\omega_n\|^2,\end{aligned}$$

hence

$$\|\omega_n\|^2 \cdot (\lambda - \tfrac{1}{2}\varrho\Lambda) \leqq \Phi(u_n) - \Phi(u_{n+1}), \tag{4.3.9}$$

therefore

$$\|G(u_n)\|^2 \leqq \frac{\Lambda^2}{\varrho(\lambda - \frac{1}{2}\varrho\Lambda)}(\Phi(u_n) - \Phi(u_{n+1})) \tag{4.3.10}$$

and we conclude the proof as that of Theorem 4.2.2.

4.3.11. COROLLARY. The global convergence of the Newton method follows if $\Lambda < 2\lambda$.

4.3.12. REMARK. Let the conditions of Theorem 4.3.6 be satisfied. Let $H(u)\,v$ be continuous in u. Then

$$\|G(u_{n+1})\| \leqq \max\left(\left|1 - \varrho\frac{\Lambda}{\lambda}\right|, \left|1 - \varrho\frac{\lambda}{\Lambda}\right|\right). \tag{4.3.13}$$

Let us consider $\varrho = 1$, the standard Newton method. Let $\omega(t) \geqq 0$ be a nondecreasing continuous function on the interval $[0, \infty)$, and such that $\omega(0) = 0$. First let us suppose that the mapping $u \to H(u)$ is continuous from H to $\mathscr{L}(H, H)$ (bounded linear operators) and that

$$\|H(u) - H(v)\| \leqq \omega(\|u - v\|). \tag{4.3.14}$$

Let us remark that, as we shall see later, the condition (4.3.14) in the weak formulation is not satisfied in general in applications to elliptic systems.

4.3.15. THEOREM. *Let the conditions of Theorem* 4.2.2 *be satisfied and let* (4.3.14) *be valid. Let* $G(u^*) = 0$ *and let*

$$\frac{1}{\lambda}\|G(u_1)\| \leqq 1, \qquad \|u^* - u_1\| \leqq 1, \tag{4.3.15$'$}$$

$$\frac{1}{\lambda}\omega\left(\frac{1}{\lambda}\|G(u_1)\|\right) \leqq \theta < 1.$$

Then we have $\|u_n - u^*\| \leqq 1$, $u_n \to u^*$ *for all the iterations* (4.3.4) *and*

$$\|G(u_{n+1})\| \leqq \omega\left(\frac{1}{\lambda}\|G(u_n)\|\right)\frac{1}{\lambda}\|G(u_n)\|, \tag{4.3.16}$$

$$\frac{1}{\lambda}\omega\left(\frac{1}{\lambda}\|G(u_n)\|\right) \leqq \theta, \tag{4.3.17}$$

$$\|u^* - u_n\| \leqq \frac{1}{\lambda}\|G(u_n)\|, \qquad \frac{1}{\lambda}\|G(u_n)\| \leqq 1. \tag{4.3.18}$$

Proof. We have $G(u_2) = G(u_1) + \left(\int_0^1 H(u_1 + t(u_2 - u_1))\,\mathrm{d}t\right)(u_2 - u_1) = \int_0^1 [H(u_1) - H(u_1 + t(u_2 - u_1))]\,\mathrm{d}t\, H^{-1}(u_1)\, G(u_1)$. We thus get in virtue of (4.3.14) that

$$\|G(u_2)\| \leqq \omega\left(\frac{1}{m}\|G(u_1)\|\right)\frac{1}{\lambda}\|G(u_1)\|. \tag{4.3.19}$$

Further, we have

$$(G(u) - G(u^*), u - u^*) \geqq \lambda \|u - u^*\|^2 \tag{4.3.20}$$

for $u \in H$, hence

$$\|u - u^*\| \leqq \frac{1}{\lambda} \|G(u)\|. \tag{4.3.21}$$

From (4.3.19), (4.3.21) the estimates of the theorem follow, especially $\|u^* - u_n\| \leqq \frac{1}{\lambda} \theta^{n-1} \|G(u_1)\|$, hence $u_n \to u^*$.

4.3.22. EXAMPLE. Consider $\Phi(u) = \frac{1}{2} \int_\Omega (|\nabla u|^2 - 2u_i f_i) \, dx + \int_\Omega F(x, u) \, dx$. Let the conditions (4.1.6) be fulfilled. We have

$$(H(u) k, h) = \int_\Omega \left[\nabla h \nabla k + \frac{\partial^2 F}{\partial \xi_r \partial \xi_s} (x, u) h_r k_s \right] dx.$$

Consider the case of $n \geqq 3$. Further suppose that

$$\left| \frac{\partial^2 F}{\partial \xi_r \partial \xi_s} (x, u) - \frac{\partial^2 F}{\partial \xi_r \partial \xi_s} (x, v) \right| \leqq c|u - v|. \tag{4.3.23}$$

The imbedding theorem implies

$$\begin{aligned} &|((H(u) - H(v)) k, h)| \\ &\leqq c\|k\|_{1,2} \|h\|_{1,2} \left(\int_\Omega \sum_{r,s=1}^m \left| \frac{\partial^2 F}{\partial \xi_r \partial \xi_s} (x, u) - \frac{\partial^2 F}{\partial \xi_r \partial \xi_s} (x, v) \right|^{\frac{n}{2}} dx \right)^{\frac{2}{n}} \\ &\leqq c\|k\|_{1,2} \|h\|_{1,2} \|u - v\|_{1,2}^\mu, \end{aligned}$$

where $\mu = 1$ for $n \leqq 6$ and $\mu = \dfrac{4}{n-2}$ for $n > 6$.

4.3.24. EXAMPLE. Let us consider $\Phi(u) = \int_\Omega (F(x, \nabla u) - u_i f_i) \, dx$, assume (4.1.6) and

$$\frac{\partial^2 F}{\partial \eta_i^r \partial \eta_j^s} (x, \eta) \zeta_i^r \zeta_j^s \geqq c|\zeta|^2. \tag{4.3.25}$$

We have

$$(H(u) k, h) = \int_\Omega \frac{\partial^2 F}{\partial \eta_i^r \partial \eta_j^s} (x, \nabla u) \frac{\partial h_r}{\partial x_i} \frac{\partial k_s}{\partial x_j} \, dx.$$

Using the Jegorov theorem and the equicontinuity of integrals with respect to the measure ($f \in S \subset L^p(\Omega)$ are equicontinuous with respect to the measure if $\forall \varepsilon > 0$, $\exists \delta > 0$ such that if $M \subset \Omega$, $|M| < \delta$, then $\forall f \in S \int_M |f|^p \, dx < \varepsilon^p$), we get that the mapping $u \mapsto H(u) k$ is continuous. It is not true that $u \mapsto H(u)$ is continuous from $[W^{1,2}(\Omega)]^m \to \mathscr{L}([W^{1,2}(\Omega)]^m, [W^{1,2}(\Omega)]^m)$ because this is equivalent to the statement: if $u_n \to u$, then

$$\lim_{n \to \infty} \sup_{\|k\|_{1,2}, \|h\|_{1,2} \leqq 1} |((H(u_n) - H(u)) k, h)| = 0 \tag{4.3.26}$$

which is not true. Clearly, if $u_n \to u$ in $[C^1(\bar{\Omega})]^m$, then (4.3.26) is true.

Let us turn back to Example 4.3.24. Suppose that $f_r \in L^p(\Omega)$, $p > n$, and let us consider, for the sake of simplicity, the Dirichlet boundary condition $u = 0$ on $\partial\Omega$.

We say that Euler's equation

$$\int_\Omega \frac{\partial F}{\partial \eta_i^r}(x, \nabla u)\frac{\partial \phi_r}{\partial x_i}\,dx = \int_\Omega f_r \phi_r\,dx, \tag{4.3.27}$$

$\forall \phi_r \in [W_0^{1,2}(\Omega)]^m$ is *regular*, if the weak solution to (4.3.27) lies in $[C^{1,\alpha}(\bar\Omega)]^m$ with $\alpha = 1 - \dfrac{n}{p}$ and if

$$\|u\|_{[C^{1,\alpha}(\bar\Omega)]^m} \leqq C(\|f\|_{[L^p(\Omega)]^m}), \tag{4.3.28}$$

where $c(s)$ is a non-decreasing bounded function on $\bar{\mathbb{R}}_+^1$.

So let $B \overset{\text{def}}{=} [W_0^{1,2}(\Omega)]^m \cap [C^{1,\alpha}(\bar\Omega)]^m$ and let us consider the iterations (4.3.4) in this space. Let $u^1 \in B$ and put $w^1 = G(u^1)$, i.e.

$$\int_\Omega \nabla w^1\,\nabla\phi\,dx = \int_\Omega \frac{\partial F}{\partial \eta_i^r}(x, \nabla u^1)\frac{\partial \phi_r}{\partial x_i}\,dx; \tag{4.3.29}$$

for simplicity we suppose that all the second derivatives of F with respect to x, η are bounded on $\bar\Omega \times \mathbb{R}^{mn}$.

From the theory of linear equations, see for example [59], (here this concerns only the Laplace operator) we get

$$\|w^1\|_{[C^{1,\alpha}(\bar\Omega)]^m} \leqq c_1(1 + \|u^1\|_{[C^{1,\alpha}(\bar\Omega)]^m}). \tag{4.3.30}$$

If $\omega = H(u)\,v$, this means that

$$\int_\Omega \nabla\omega\,\nabla\phi\,dx = \int_\Omega \frac{\partial^2 F}{\partial \eta_i^r\,\partial \eta_j^s}(x, \nabla u)\frac{\partial v_s}{\partial x_j}\,\frac{\partial \phi_r}{\partial x_i}\,dx. \tag{4.3.31}$$

Once more, it follows from the theory of linear equations that

$$\|H^{-1}(u)\|_{\mathscr{L}([C^{1,\alpha}(\bar\Omega)]^m,\,[C^{1,\alpha}(\bar\Omega)]^m)} \leqq c_2(\|u\|_{[C^{1,\alpha}(\bar\Omega)]^m}), \tag{4.3.32}$$

where c_2 is another bounded function on $\bar{\mathbb{R}}_+^1$. If we moreover suppose that all the third derivatives of F with respect to x and η are bounded on $\bar\Omega \times \mathbb{R}^{mn}$, then we get easily from (4.3.31), using only the regularity for the Laplace operator, that

$$\begin{aligned} &\|H(u) - H(v)\|_{\mathscr{L}([C^{1,\alpha}(\bar\Omega)]^m,\,[C^{1,\alpha}(\bar\Omega)]^m)} \\ &\leqq c_3(\|u\|_{[C^{1,\alpha}(\bar\Omega)]^m}, \|v\|_{[C^{1,\alpha}(\bar\Omega)]^m})\,\|u - v\|_{[C^{1,\alpha}(\bar\Omega)]^m}, \end{aligned} \tag{4.3.33}$$

where c_3 is a bounded function on $\bar{\mathbb{R}}_+^2$.

I believe that it is not necessary to persuade the reader of some complications that the Newton method brings; especially (4.3.28) is very delicate and we shall discuss it in Chapter 6.

The abstract formulation is as follows: suppose that all the conditions of Theorem 4.3.6 are satisfied. Let B be a Banach space $B \subsetneq H$, let $G(u^*) = 0$ and let $u^* \in B$. Let

(4.3.34) $\qquad u \mapsto G(u)$ be a continuous map from B to B;

denote by (,) and | | the scalar product and the norm in H, respectively, by $\| \ \|$ the norm in B, and let $H(u) \in \mathscr{L}(B, B)$ be such that

$$\|H(u) - H(v)\| \leqq f(\|u\| + \|v\|)\,\omega(\|u - v\|), \tag{4.3.35}$$

where f is a bounded non-decreasing function in $\overline{\mathbb{R}}^1_+$ and ω is a continuous non-decreasing function in $\overline{\mathbb{R}}^1_+$ such that $\omega(0) = 0$. Suppose moreover that

$$\max_{\|u - u\|^* \leqq \delta} \|H^{-1}(u)\| = \alpha(\delta). \tag{4.3.36}$$

4.3.37. THEOREM. *Let the conditions* (4.3.34)–(4.3.36) *be satisfied. Put* $\beta(\delta) = f(4\delta + 2\|u^*\|)$ *and suppose*

$$\alpha(\delta)\,\beta(\delta)\,\omega(\delta) < 1. \tag{4.3.38}$$

Put $A(u) = u - H^{-1}(u)\,G(u)$. *Then*

$$\|Au - u^*\| \leqq \delta \tag{4.3.39}$$

for $\|u - u^*\| \leqq \delta$. *Further suppose that*

$$\alpha(\delta)\,\beta(\delta)\,\omega(\alpha(\delta)\,\|G(u_1)\|) \leqq \theta < 1. \tag{4.3.40}$$

Then

$$\|G(u_{n+1})\| \leqq \alpha(\delta)\,\beta(\delta)\,\omega(\alpha(\delta)\,\|G(u_n)\|)\,\|G(u_n)\|, \tag{4.3.41}$$

holds for the iterations (4.3.4), *especially*

$$\|G(u_{n+1})\| \leqq \theta\|G(u_n)\| \tag{4.3.42}$$

and

$$\|u_n - u^*\| \leqq \frac{\alpha(\delta)}{1 - \alpha(\delta)\,\beta(\delta)\,\omega(\delta)}\,\|G(u_n)\|. \tag{4.3.43}$$

Proof. Let us first consider (4.3.39). We have

$$Au - u^* = H^{-1}(u)\left[H(u)(u - u^*) - \left(\int_0^1 H(u^* + t(u - u^*))\,dt\right)(u - u^*)\right], \tag{4.3.44}$$

hence

$$\|Au - u^*\| \leqq \alpha(\delta)\,\beta(\delta)\,\omega(\delta)\,\delta. \tag{4.3.45}$$

Further, we have

$$G(u_2) = \int_0^1 (H(u_1) - H(u_1 + t(u_2 - u_1)))\,dt \cdot H^{-1}(u_1)\,G(u_1), \tag{4.3.46}$$

hence

$$\|G(u_2)\| \leqq \beta(\delta)\,\omega(\alpha(\delta)\,\|G(u_1)\|)\,\alpha(\delta)\,\|G(u_1)\|. \tag{4.3.47}$$

Now

$$G(u_n) - G(u^*) = \left(\int_0^1 [H(u^* + t(u_n - u^*)) - H(u_n)]\,dt\right)(u_n - u^*) + H(u_n)(u_n - u^*). \tag{4.3.48}$$

But (4.3.43) follows from (4.3.48) immediately.

Let us remark once more that the potentiality does not play any role in all the theorems of this section.

4.3.44. REMARK. If we have $\|H(u) - H(v)\| \leqq c(1 + \|u\| + \|v\|)\,\|u - v\|$ instead of (4.3.35) then

$$\|G(u_n)\| \leqq \lambda \|G(u_n)\|^2 \tag{4.3.45}$$

and

$$\|G(u_n)\| \leqq \frac{1}{\lambda}(\lambda \|G(u_1)\|)^{2^{n-1}}. \tag{4.3.46}$$

4.4. Differentiable homotopy

We consider, as above, because of a uniform point of view, a potential Φ from $H \to \mathbb{R}^1$, twice differentiable, with $H(u)$ such that $u \mapsto H(u)\,v$ is a continuous mapping; we suppose, as usual, that

$$|D^2\Phi(u, h, h)| \leqq \Lambda \|h\|^2, \tag{4.4.1}$$

$$D^2\Phi(u, h, h) \geqq \lambda \|h\|^2. \tag{4.4.2}$$

Assume the functional in the form $\Phi(u) = \Psi(u) - (f, u)$, where $f \in H$; let $f = f(t)$, $t \in [0, 1]$ and $f \in C^1([0, 1], H)$.

4.4.3. THEOREM. *Let the conditions mentioned above be satisfied. Then for $u(t)$, satisfying $G(u(t)) = 0$, we have $u \in C^1([0, 1], H)$ and*

$$H(u(t))\,u'(t) = f'(t). \tag{4.4.4}$$

Proof. We have, as usual, that

$$\begin{aligned} &D\Psi(u(t + \Delta t), h) - D\Psi(u(t), h) \\ &= \int_0^1 D^2\Psi(u(t) + \tau(u(t + \Delta t) - u(t)), h, u(t + \Delta t) - u(t))\,d\tau \\ &= (f(t + \Delta t) - f(t), h). \end{aligned} \tag{4.4.5}$$

Put $h = u(t + \Delta t) - u(t)$. We see that

$$\lambda \|u(t + \Delta t) - u(t)\| \leqq \|f(t + \Delta t) - f(t)\|, \tag{4.4.6}$$

hence $u \in C([0, 1], H)$.

Let $\Delta t \to 0$. Since (4.4.6) implies

$$\lambda \left\| \frac{\Delta u}{\Delta t} \right\| \leqq \left\| \frac{\Delta f}{\Delta t} \right\|, \tag{4.4.7}$$

if $\Delta t = \lambda_n \to 0$, we can suppose $\dfrac{\Delta u}{\Delta t} \rightharpoonup \omega(t)$. But it follows from (4.4.5) that

$$(H(u(t))\,\omega(t), h) = (f'(t), h), \tag{4.4.8}$$

hence $\omega(t)$ is uniquely defined by (4.4.8) and thus $\dfrac{\Delta u}{\Delta t} \rightharpoonup \omega$ for $\Delta t \to 0$. Let us prove that $\dfrac{\Delta u}{\Delta t} \to \omega$. Put $\int_0^1 H(u(t) + \tau(u(t + \Delta t) - u(t)))\,d\tau \cdot k = H_{\Delta t}(u(t))\,k$.

We thus get from (4.4.5) and (4.4.8) that

$$\left(H_{\Delta t}(u(t))\left(\frac{\Delta u}{\Delta t}(t) - \omega(t)\right), \frac{\Delta u}{\Delta t}(t) - \omega(t)\right) = \left(\frac{\Delta f}{\Delta t}(t) - f'(t), \frac{\Delta u}{\Delta t}(t) - \omega(t)\right) - \left((H_{\Delta t}(u(t)) - H(u(t)))\,\omega(t), \frac{\Delta u}{\Delta t}(t) - \omega(t)\right), \tag{4.4.9}$$

from which $\frac{\Delta u}{\Delta t}(t) \to \omega(t) = u'(t)$ follows. The same trick leads to the continuity of $u'(t)$.

4.4.10. THEOREM. *Let the conditions of Theorems* 4.1.10 *and* 4.4.3 *be satisfied. Let $u_n(t)$ be the minimum of the functional $\Phi(v)$ over H_n. Then $u_n \in C^1([0, 1], H)$, for every $v \in H_n$*

$$(H(u_n(t))\, u_n'(t), v) = (f'(t), v) \tag{4.4.11}$$

and $u_n \to u$ in $C^1([0, 1], H)$.

Proof. Clearly (4.4.11) can be obtained as in the proof of the preceding theorem. Of course, since $\dim H_n = n$, the weak and strong convergence coincide.

It follows from (4.4.11) that

$$\|u_n'\|_{C([0,1], H)} \leqq \frac{1}{\lambda} \|f'\|_{C([0,1], H)} \tag{4.4.12}$$

and from (4.4.4) that

$$\|u'\|_{C([0,1], H)} \leqq \frac{1}{\lambda} \|f'\|_{C([0,1], H)}. \tag{4.4.13}$$

Now

$$\Phi(u_n(t + \Delta t)) - \Phi(u_n(t)) - D\Phi(u_n(t), u_n(t + \Delta t) - u_n(t)) = \int_0^1\int_0^1 D^2\Phi(u_n(t) + s\sigma(u_n(t + \Delta t) - u_n(t)), u_n(t + \Delta t) - u_n(t), u_n(t + \Delta t) - u_n(t))\, s \,ds\, d\sigma, \tag{4.4.14}$$

hence

$$|\Phi(u_n(t + \Delta t)) - \Phi(u_n(t))| \leqq \frac{\Lambda}{2}\frac{1}{\lambda^2} \|f'\|^2_{C([0,1], H)} |\Delta t|^2. \tag{4.4.15}$$

In the same way we obtain

$$|\Phi(u(t + \Delta t)) - \Phi(u(t))| \leqq \frac{\Lambda}{2\lambda^2} \|f'\|_{C([0,1], H)} (\Delta t)^2. \tag{4.4.16}$$

Let $\varepsilon > 0$. Choose $|\Delta t|$ so small that

$$\frac{\Lambda}{2\lambda^2} \|f'\|^2_{C([0,1], H)} |\Delta t|^2 < \frac{\varepsilon}{3}. \tag{4.4.17}$$

Let $t_k = k\,\Delta t$, $k = 1, 2, \ldots, M$, and choose n so large that

$$\Phi(u_n(t_k)) \leqq \Phi(u(t_k)) + \frac{\varepsilon}{3}. \tag{4.4.18}$$

It follows from (4.4.15)–(4.4.18) that

$$\Phi(u_n(t)) \leqq \Phi(u(t)) + \varepsilon \tag{4.4.19}$$

and (4.1.17) implies $u_n \to u$ in $C([0, 1], H)$. Let now P_n be the orthogonal projector onto H_n. Clearly

$$\lim_{n\to\infty} \max_{[0,1]} \|P_n u'(t) - u'(t)\| = 0. \tag{4.4.20}$$

We have

$$(H(u_n(t)) - H(u(t)))\, u'(t) \to 0 \tag{4.4.21}$$

uniformly in [0, 1]. Now

$$\begin{aligned}(H(u_n(t))\,(u'(t) - u_n'(t)), u'(t) - u_n'(t))\\ = (H(u_n(t))\,(P_n u'(t) - u_n'(t)), P_n u'(t) - u_n'(t)) + \varepsilon_n(t),\end{aligned} \tag{4.4.22}$$

where $\varepsilon_n(t) \to 0$ uniformly in [0, 1]. On the other hand,

$$\begin{aligned}(H(u_n)\,(P_n u'(t) - u_n'(t)), P_n u'(t) - u_n'(t))\\ = -(f'(t), P_n u'(t) - u_n'(t)) + (H(u_n(t))\, P_n u'(t),\end{aligned} \tag{4.4.23}$$

$$P_n u'(t) - u_n'(t)) = -(f'(t), u'(t) - u_n'(t)) + (H(u_n(t))\, u'(t), u'(t) - u_n'(t)) + \bar{\varepsilon}_n(t)$$

with $\bar{\varepsilon}_n(t) \to 0$ uniformly in [0, 1]. Taking into account (4.4.21), we get from (4.4.23) that $-(f'(t), u'(t) - u_n'(t)) + (H(u_n(t))\,u'(t), u'(t) - u_n'(t)) + \bar{\varepsilon}_n(t) = \tilde{\bar{\varepsilon}}_n(t)$ with $\tilde{\bar{\varepsilon}}_n(t) \to 0$ uniformly in [0,1], thus $u_n'(t) \to u'(t)$ uniformly in [0, 1].

The simplest linearization of the differential equation $H(u(t))\,u'(t) = f'(t)$, $u(0) = u_0$, is the Euler method: let $t_k = k\,\Delta t$, $k = 0, 1, 2, \ldots, M$, $M\,\Delta t = 1$, and put

$$\tilde{u}(t_{k+1}) = \tilde{u}(t_k) + \Delta t H^{-1}(\tilde{u}(t_k)) f'(t_k). \tag{4.4.24}$$

We suppose, of course, that the function $f(t)$ is chosen in such a way that the solution to $\Psi'(u(0)) = f(0)$ is known, for example, $u(0) = 0$. For $t_k \leqq t \leqq t_{k+1}$, $k = 0, 1, 2, \ldots, M - 1$, we put $u_M(t) \overset{\text{def}}{=} \tilde{u}(t_k)\left(1 - \dfrac{t - t_k}{\Delta t}\right) + \tilde{u}(t_{k+1})\dfrac{t - t_k}{\Delta t}$. Clearly, we ask, in what sense (if in any) $u_M(t) \to u(t)$. As in the Newton method, we need some sort of regularity, but much less than in the case of this method, and we shall see in an example that this regularity is satisfied in general. In general, this is an open:

4.4.25. PROBLEM. Under the conditions of Theorem 4.4.3, is $u_M(t) \to u(t)$ in some sense?

We proceed with an abstract theorem; first let us formulate the regularity requirement: Let $B \subsetneq H$ be a Banach space such that $H^{-1}(u)$ maps B into itself,

$$\|H^{-1}(u)\, h\|_B \leqq c_1 \|h\|_B, \tag{4.4.26}$$

and

(4.4.27) $$\|(H(u+v)-H(u))\,g\|_H \leqq c_2(\|u\|_H+\|g\|_B)\,\omega(\|v\|_H),$$

where c_2 is a non-decreasing finite function, ω is a non-decreasing continuous finite function such that $\omega(0)=0$.

4.4.28. THEOREM. *Let the conditions of Theorem* 4.4.3 *be satisfied. Let* (4.4.26) *and* (4.4.27) *be valid and let* $f \in C^1([0,1], B)$. *Then* $u_M \to u$ *in* $C([0,1], H)$, $u'_{M+}(t) \to u'(t)$ *in* H *and uniformly in* $[0,1]$.

4.4.29. EXAMPLE. Consider the functional from Example 4.3.24 and assume (4.1.6), (4.3.25). We have $H = [W_0^{1,2}(\Omega)]^m$ and put $B \overset{\text{def}}{=} [W_0^{1,p}(\Omega)]^m$ with $2 < p$; p will be fixed later. Suppose that $\partial\Omega$ is smooth enough. Suppose that all the third derivatives of F with respect to η are bounded in $\bar{\Omega} \times \mathbb{R}^{mn}$. First, we have (4.4.27). Let $\|u-v\|_{1,2} \leqq \delta$. Let $M_\delta = \{x \in \Omega \mid |\nabla u(x) - \nabla v(x)| > \sqrt{\delta}\}$. We have $|M_\delta| \leqq \delta$. For $k \in [W_0^{1,2}(\Omega)]^m$, $g \in [W_p^{1,2}(\Omega)]^m$, it is

$$\int_\Omega \left(\frac{\partial^2 F}{\partial\eta_i^r\,\partial\eta_j^s}(x,\nabla u) - \frac{\partial^2 F}{\partial\eta_i^r\,\partial\eta_j^s}(x,\nabla v)\right)\frac{\partial g_r}{\partial x_i}\frac{\partial k_s}{\partial x_j}\,dx = \int_{\Omega\setminus M_\delta} + \int_{M_\delta}.$$

But

(4.4.30) $$\left|\int_{\Omega\setminus M_\delta}\right| = \left|\int_{\Omega\setminus M_\delta}\left(\int_0^1 \frac{\partial^3 F}{\partial\eta_i^r\,\partial\eta_j^s\,\partial\eta_k^t}(x,\nabla v + t(\nabla u - \nabla v))\,dt\right)\right.$$
$$\left.\times\left(\frac{\partial u_t}{\partial x_k} - \frac{\partial v_t}{\partial x_k}\right)\frac{\partial g_r}{\partial x_i}\frac{\partial k_s}{\partial x_j}\,dx\right| \leqq c_1\sqrt{\delta}\,\|g\|_{1,2}\,\|k\|_{1,2},$$

(4.4.31) $$\left|\int_{M_\delta}\right| \leqq c_2\left(\int_{M_\delta}|\nabla g|^2\,dx\right)^{\frac{1}{2}}\|k\|_{1,2} \leqq c_2\delta^{\frac{p-2}{2p}}\|g\|_{1,p}\|k\|_{1,2},$$

hence (4.4.27) follows.

Further let $f \in [W_0^{1,p}(\Omega)]^m$ and let us look for $h \in [W_0^{1,p}(\Omega)]^m$ such that

(4.4.32) $$\int_\Omega \frac{\partial^2 F}{\partial\eta_i^r\,\partial\eta_j^s}(x,\nabla u)\frac{\partial h_r}{\partial x_i}\frac{\partial k_s}{\partial x_j}\,dx = \int_\Omega \frac{\partial f_r}{\partial x_i}\frac{\partial k_r}{\partial x_i}\,dx$$

holds $\forall k \in [W_0^{1,p^*}(\Omega)]^m$. Due to the result of N. G. Meyers [44] or J. Nečas [45], there exists $p > 2$, $p = p\left(\frac{\lambda_1}{\lambda_2}\right)$, where $\lambda_1|\xi|^2 \leqq \frac{\partial^2 F}{\partial\zeta_i^r\partial\zeta_j^s}(x,\zeta)\,\xi_i^r\xi_j^s \leqq \lambda_2|\xi|^2$, such that such a solution h to (4.4.32) exists and

(4.4.33) $$\|h\|_{1,p} \leqq c_3\|f\|_{1,p}.$$

This is (4.4.26). Let us remark, that if $\Omega = T(O)$, where T is a bi-Lipschitz homeomorphism of O onto Ω and O has ∂O smooth enough, then, considering $k(T(y))$, we get also the result (4.4.33).

Proof of Theorem 4.4.28: First we have

$$\|u'_{M+}(t)\| \leqq \lambda^{-1}\|f'\|. \tag{4.4.34}$$

Since

$$H^{-1}(u) - H^{-1}(v) = -H^{-1}(u)(H(u) - H(v))H^{-1}(v) \tag{4.4.34'}$$

we get

$$\begin{aligned} &\|H^{-1}(\tilde{u}(t_k))f'(t_k) - H^{-1}(u_M(t))f'(t)\| \\ &\leqq \frac{1}{\lambda}\|f'(t_k) - f'(t)\| + c_1\omega(c_2|\Delta t|) \end{aligned} \tag{4.4.35}$$

for $t_k \leqq t \leqq t_{k+1}$. Hence

$$u'_{M+}(t) = H^{-1}(u_M(t))f'(t) + g_M(t), \tag{4.4.36}$$

where $g_M \to 0$ in $C([0, 1], H)$. Therefore

$$\Psi'(u_M(t)) = f(t) + \int_0^t g_M(\tau)\,d\tau. \tag{4.4.37}$$

Hence $u_M \to u$ in $C([0, 1], H)$. By (4.4.34)′ (approximating $f(t)$ by piecewise constant functions in $[0, 1]$), we have $\lim_{M\to\infty}(H^{-1}(u_M(t)) - H^{-1}(u(t)))f'(t) = 0$ in $C([0, 1], H)$, hence the result follows.

Considering the method of successive approximations

$$u_{n+1}(t) = u_0 + \int_0^t H^{-1}(u_n(\tau))f'(\tau)\,d\tau, \tag{4.4.38}$$

we come to the same troubles as in Theorem 4.3.37, i.e. it seems that we must assume the conditions (4.3.34)–(4.3.36). We assume then in a little stronger form.

4.4.39. THEOREM. *Let the conditions of Theorem* 4.4.3 *be satisfied. Let there exist a Banach space* $B \subsetneq H$ *such that*

$$H(u) \in \mathscr{L}(B, B) \quad \text{for } u \in B, \tag{4.4.40}$$

$$\|H^{-1}(u)\|_B \leqq c_1(1 + \|u\|_B), \quad u \in B, \tag{4.4.41}$$

$$\|H(u) - H(v)\|_B \leqq c_2(1 + \|u\|_B + \|v\|_B)\|u - v\|_B. \tag{4.4.42}$$

Let $f \in C^1([0, 1], B)$, $u_0 \in B$. *Then there exists a unique solution* $u \in C^1([0, 1], B)$, $u(0) = u_0$, *to the equation*

$$H(u(t))\,u'(t) = f'(t) \tag{4.4.43}$$

and $\Delta t > 0$ *such that the method of successive approximations*

$$u_{n+1}(t) = u(t_0) + \int_{t_0}^t H^{-1}(u_n(\tau))f'(\tau)\,d\tau \tag{4.4.44}$$

converges for $|t - t_0| \leqq \Delta t$.

Proof. This is a classical situation, let us remark only that (4.4.41) implies

(4.4.45) $$\|u(t)\|_B \leqq \|u(t_0)\|_B + c_1\Big(1 + \int_{t_0}^{t} \|u(\tau)\|_B \, d\tau\Big),$$

from which $\|u(t)\|_B \leqq c_3$ follows "a priori", hence the global result as well.

4.4.46. EXERCISE. Prove under the conditions of Theorem 4.4.39 that the method of successive approximations (4.4.38) converges in $C([0, 1], B)$ in the norms $\max_{[0,1]} \|u(t)\, e^{-\alpha t}\|_B$, where $\alpha > 0$ is large enough.

4.5. Secant modulus method

This method is described in more detail in the book [19] and in the lecture notes [7] where also other references can be found. In this chapter we follow the work by J. NEČAS [46]. Let us begin with an example.

4.5.1. EXAMPLE. For the sake of simplicity, look for the functional $\Phi(u) = \int_\Omega |F(\nabla u) - u_i f_i| \, dx$ with the standard conditions: $f \in [L^2(\Omega)]^m$, $\sum_{r\,i,s,j} \left|\frac{\partial^2 F}{\partial \eta_i^r \partial \eta_j^s}\right| \leqq c$, and $\frac{\partial^2 F}{\partial \eta_i^r \partial \eta_j^s}(\eta)\, \zeta_i^r \zeta_j^s \geqq c|\zeta|^2$. Let us suppose especially that

(4.5.2) $$\frac{\partial F}{\partial \eta_i^r}(\nabla u) = A_{ij}^{rs}(\nabla u)\frac{\partial u_s}{\partial x_j}$$

and let $A_{ij}^{rs}(\eta) = A_{ji}^{sr}(\eta)$; let $A_{ij}^{rs}(\eta)$ be continuous functions in $\mathbb{R}^{mn}$ and let

(4.5.3) $$\lambda_1|\zeta|^2 \leqq A_{ij}^{rs}(\eta)\, \zeta_i^r \zeta_j^s \leqq \lambda_2|\zeta|^2.$$

The critical point of the functional Φ satisfies the equation

(4.5.4) $$\int_\Omega A_{ij}^{rs}(\nabla u)\frac{\partial \phi_r}{\partial x_i}\frac{\partial u_s}{\partial x_j}\, dx = \int_\Omega \phi_r f_r \, dx$$

$\forall \phi \in [W_0^{1,2}(\Omega)]^m$; let us look for $u \in [W_0^{1,2}(\Omega)]^m$.

The secant modulus method is defined as follows: Let u^n be the n-th approximation. The approximation u^{n+1} is the solution to the linear problem

(4.5.5) $$\int_\Omega \left[A_{ij}^{rs}(\nabla u^n)\frac{\partial \phi_r}{\partial x_i}\frac{\partial u_s^{n+1}}{\partial x_j} - \phi_r f_r\right] dx = 0,$$

which is equivalent to the search for the minimum of the functional

$$\frac{1}{2}\int_\Omega \left[A_{ij}^{rs}(\nabla u^n)\frac{\partial u_r}{\partial x_i}\frac{\partial u_s}{\partial x_j} - 2u_r f_r\right] dx$$

in $[W_0^{1,2}(\Omega)]^m$. It turns out that the fundamental condition for the convergence of this method is

(4.5.6) $$\Phi(u_{n+1}) \leqq \Phi(u_n).$$

The defined iterations have often a clear physical meaning (see [19]) and the method leads to a very rapidly convergent process in practice.

4.5.7. THEOREM. *Let the functional Φ be defined on a Hilbert space H and, as usual, let*

$$D\Phi(u + h, h) - D\Phi(u, h) \geqq c_1\|h\|^2, \quad c_1 > 0.$$

Let there exist a bilinear form $B(u; x, y)$ in x, y such that

$$B(u; x, x) \geqq c_2\|x\|^2, \quad c_2 > 0, \tag{4.5.8}$$

$$|B(u; x, y)| \leqq c_3\|x\| \cdot \|y\|, \tag{4.5.9}$$

$$B(u; u, v) = D\Phi(u, v) \tag{4.5.10}$$

$$\tfrac{1}{2}B(x; y, y) - \tfrac{1}{2}B(x; x, x) - \Phi(y) + \Phi(x) \geqq 0. \tag{4.5.11}$$

Let $\phi \in H$ and let u be a solution to the problem

$$D\Phi(u, v) = (\phi, v), \quad \forall v \in H. \tag{4.5.12}$$

If u_n is the n-th iterate, define u_{n+1} as

$$B(u_n; u_{n+1}, v) = (\phi, v), \quad \forall v \in H. \tag{4.5.13}$$

Then $u_n \to u$ in H (clearly both u and u_n are uniquely determined).

Proof. Put

$$\pi_n(v) = \Phi(u_n) - (v, \phi) + \tfrac{1}{2}B(u_n; v, v) - \tfrac{1}{2}B(u_n; u_n, u_n), \tag{4.5.14}$$

$$\Psi(v) = \Phi(v) - (v, \phi). \tag{4.5.15}$$

The relation (4.5.11) implies

$$\Psi(u_{n+1}) \leqq \pi(u_{n+1}) \leqq \pi_n(u_n) = \Psi(u_n). \tag{4.5.16}$$

Further it follows from (4.5.8) that

$$\begin{aligned} c_2\|u_{n+1} - u_n\|^2 &\leqq B(u_n; u_{n+1} - u_n, u_{n+1} - u_n) \\ &= (\phi, u_{n+1} - 2u_n) + B(u_n; u_n, u_n) = 2(\Psi(u_n) - \pi_n(u_{n+1})) \\ &\leqq 2(\Psi(u_n) - \Psi(u_{n+1})), \end{aligned} \tag{4.5.17}$$

hence ($\Psi(u)$ is bounded from below)

$$\lim_{n\to\infty} \|u_{n+1} - u_n\| = 0, \tag{4.5.18}$$

$$\begin{aligned} c_1\|u_n - u\|^2 &\leqq D\Phi(u_n, u_n - u) - D\Phi(u, u_n - u) \\ &= B(u_n; u_n, u_n - u) - (\phi, u_n - u) = B(u_n; u_{n+1}, u_n - u) \\ &\quad - (\phi, u_n - u) + B(u_n; u_n - u_{n+1}, u_n - u_{n+1}) \\ &= B(u_n; u_n - u_{n+1}, u_n - u) \leqq c_3\|u_n - u_{n+1}\| \cdot \|u_n - u\|, \end{aligned} \tag{4.5.19}$$

therefore the result follows from (4.5.19) and (4.5.18).

4.5.20. THEOREM. *Consider Example* 4.5.1. *Let us suppose that*

$$\frac{\partial F}{\partial \eta_i^r}(\xi)\,(\eta_i^r - \xi_i^r) + F(\xi) - F(\eta) + \frac{1}{2}A_{ij}^{rs}(\xi)\,(\eta_i^r - \xi_i^r)\,(\eta_j^s - \xi_j^s) \geqq 0. \tag{4.5.21}$$

If we put

$$B(u, u, w) = \int_{\Omega} A_{ij}^{rs}(\nabla u) \frac{\partial v_r}{\partial x_i} \frac{\partial w_s}{\partial x_j} \, dx \tag{4.5.22}$$

then the conditions of Theorem 4.5.7 are fulfilled.

The proof is left to the reader.

4.5.23. REMARK. The condition (4.5.21) is clearly equivalent to the condition

$$\left[A_{ij}^{rs}(\xi) - \frac{\partial^2 F}{\partial \eta_i^r \partial \eta_j^s}(\xi + \theta h)\right] h_i^r h_j^s \geqq 0, \tag{4.5.24}$$
$$\xi, h \in \mathbb{R}^{mn}, \quad 0 < \theta < 1.$$

4.5.25. EXAMPLE. Let $\Phi(u) = \frac{1}{2} \int_{\Omega} F(|\nabla u|^2) \, dx$. Let $F \in C^2(\overline{\mathbb{R}}_+^1)$ and let

$$0 < c_1 \leqq F'(s) \leqq c_2, \tag{4.5.26}$$

$$0 < c_3 \leqq F'(s) + 2sF''(s) \leqq c_4. \tag{4.5.27}$$

Then $|D^2\Phi(u, h, k)| \leqq c_5 \|h\| \, \|k\|$, (for $u, h, k \in [W_0^{1,2}(\Omega)]^m$, $\|h\|^2 = \int_{\Omega} |\nabla h|^2 \, dx$), $D^2\Phi(u, h, h) \geqq c_6 \|h\|^2$.

If

$$F''(s) \leqq 0, \tag{4.5.28}$$

then all the conditions of Theorem 4.5.7 are fulfilled, provided that

$$B(u, v, w) \overset{\text{def}}{=} \int_{\Omega} 2F'(|\nabla u|^2) \, \nabla v \cdot \nabla w \, dx.$$

Chapter 5

Intermediary regularity

5.1. Introductory remarks

All the time we consider our second order systems

$$-\frac{\partial}{\partial x_i}[a_i^r(x, u, \nabla u)] + a^r(x, u, \nabla u) = f^r, \quad r = 1, 2, \ldots, m. \tag{5.1.1}$$

If we consider weak solutions in $[W^{1,2}(\Omega)]^m$ (under suitable growth conditions, see Chapter 3), then the standard ellipticity conditions

$$\frac{\partial a_i^r}{\partial \eta_j^s}(x, \xi, \eta)\, \zeta_i^r \zeta_j^s \geqq c|\zeta|^2, \quad c > 0, \tag{5.1.2}$$

and natural regularity conditions for the coefficients and data imply that $u \in [W_{\text{loc}}^{2,2}(\Omega)]^m$ or $u \in [W^{2,2}(\Omega)]^m$. By the same argument as mentioned in Example 4.4.29, we get $u \in [W_{\text{loc}}^{2,p}(\Omega)]^m$ or $[W^{2,p}(\Omega)]^m$ with $p > 2$. In general, for $n \geqq 3$, it is not $p > n$, thus the regularity $u \in [C^{1,\mu}(\Omega)]^m$ or $[C^{1,\mu}(\bar{\Omega})]^m$ is not valid and it cannot be obtained without further assumptions, for example, without the Liouville type assumptions, as mentioned in 1.1.

Besides the $W^{2,2}$ regularity, we shall be interested in this chapter in the inclusions $u \in [W_{\text{loc}}^{1,\infty}(\Omega)]^m$ or $u \in [W^{1,\infty}(\Omega)]^m$ which follow from some asymptotic properties of the coefficients $\frac{\partial a_i^r}{\partial \eta_j^s}$ in η. If $m = 1$, then such estimates follow from the maximum principle. In this case, we get a theorem of the Harnack type that implies the $C^{1,\mu}$ regularity. To this type of conditions also the bounded slope condition belongs, that gives a $W^{1,\infty}$ estimate up to the boundary. We shall repeat the result from the work by G. Stampacchia [47] here.

5.2. Estimates in the space $W^{2,2}$

5.2.1. LEMMA. *Let Ω be a domain with a Lipschitz boundary and let $\Omega_h = \{x \in \Omega \mid \operatorname{dist}(x, \partial\Omega) > h\}$. Let $u \in W^{1,p}(\Omega)$, $1 \leqq p \leqq \infty$, and put*

$$\Delta_l^\tau u(x) \overset{\text{def}}{=} \tau^{-1}[u(x + \tau\, e^l) - u(x)], \tag{5.2.2}$$

$e^l = (0, \dots 0, 1, 0, \dots 0)$ *with a* 1 *in the* l-th *position. For* $|\tau| \leqq h$ *then*

$$\|\Delta_l^\tau u\|_{L^p(\Omega_h)} \leqq \left\| \frac{\partial u}{\partial x_l} \right\|_{L^p(\Omega)}. \tag{5.2.3}$$

Proof. It is sufficient to consider the case of $p < \infty$. According to Theorem 2.1.9, $C^1(\bar{\Omega})$ is a dense subset of $W^{1,p}(\Omega)$, so it is enough to prove (5.2.3) for $u \in C^1(\bar{\Omega})$. But

$$\int_{\Omega_h} |\Delta_h^\tau u|^p \,\mathrm{d}x \leqq \int_{\Omega_h} \mathrm{d}x \left| \int_0^1 \frac{\partial u}{\partial x_l}(x + t\tau\, e^l)\,\mathrm{d}t \right|^p \leqq \int_{\Omega_h} \mathrm{d}x \int_0^1 \left| \frac{\partial u}{\partial x_l}(x + t\tau\, e^l) \right|^p \mathrm{d}t$$

$$\leqq \int_0^1 \mathrm{d}t \int_\Omega \left| \frac{\partial u}{\partial x_l}(y) \right|^p \mathrm{d}y.$$

5.2.4. LEMMA. *Let* $u \in L^p(\Omega)$, $1 < p \leqq \infty$, *let* $\Delta_l^\tau u \in L^p(\Omega_h)$ $\forall h > 0$, $|\tau| < h$, *and let* $\|\Delta_l^\tau u\|_{L^p(\Omega_h)} \leqq c_1 < \infty$. *For* $\dfrac{\partial u}{\partial x_l}$, *in the sense of distributions, then*

$$\left\| \frac{\partial u}{\partial x_l} \right\|_{L^p(\Omega)} \leqq c_1. \tag{5.2.5}$$

Proof. For $\phi \in \mathscr{D}(\Omega)$ and $|\tau|$ small enough we have

$$\int_\Omega \Delta_l^\tau u \phi \,\mathrm{d}x = - \int_\Omega u\, \Delta^{-\tau}\, e^l \phi \,\mathrm{d}x. \tag{5.2.6}$$

Let $p < \infty$, let $\tau_n \to 0$ in such a way that $\Delta_l^{\tau_n} u \rightharpoonup w$ in $L^p(\Omega')$ $\forall \Omega' \subset \bar{\Omega}' \subset \Omega$. It follows from (5.1.6) that $\displaystyle\int_\Omega w\phi \,\mathrm{d}x = - \int_\Omega u \frac{\partial \phi}{\partial x_l} \mathrm{d}x$.

Let $u \in [W^{1,2}(\Omega)]^m$ be a weak solution to the system

$$-\frac{\partial}{\partial x_i}[a_i^r(x, u, \nabla u)] + a^r(x, u, \nabla u) = f^r, \quad r = 1, 2, \dots, m, \tag{5.2.7}$$

and let $f^r \in L^2(\Omega)$. Let the coefficients a_i^r, a^r be once continuously differentiable with respect to all their arguments in $\bar{\Omega} \times \mathbb{R}^m \times \mathbb{R}^{mn}$ and let

$$\sum_{r,i} |a_i^r| + \sum_r |a^r| + \sum_{r,i,l} \left| \frac{\partial a_i^r}{\partial x_l} \right| + \sum_{r,i} \left| \frac{\partial a^r}{\partial x_i} \right| \leqq c(|\xi| + |\eta|), \tag{5.2.8}$$

$$\sum_{r,i,s,j} \left| \frac{\partial a_i^r}{\partial \eta_j^s} \right| + \sum_{r,i,s} \left| \frac{\partial a_i^r}{\partial \xi_s} \right| + \sum_{r,s,j} \left| \frac{\partial a^r}{\partial \eta_j^s} \right| + \sum_{r,s} \left| \frac{\partial a^r}{\partial \xi_s} \right| \leqq c, \tag{5.2.9}$$

$$\frac{\partial a_i^r}{\partial \eta_j^s}(x, \xi, \eta)\, \zeta_i^r \zeta_j^s \geqq c|\zeta|^2, \quad c > 0. \tag{5.2.10}$$

Let $\Omega \subset \mathbb{R}^n$ be a domain with a Lipschitz boundary and let $\sigma \in W_0^{1,\infty}(\Omega) \cap C^\infty(\Omega)$, $\sigma \geqq 0$. Let us suppose that there exists $\sigma_h \in \mathscr{D}(\Omega)$, $\sigma_h \geqq 0$, such that $\sigma_h \to \sigma$ in $C(\bar{\Omega})$ as $h \to 0$ $\|\sigma_h\|_{W^{1,\infty}(\Omega)} \leqq c_1$, and $\sigma_h(x) \leqq \sigma(x)$ in Ω.

5.2.11. THEOREM. *Let Ω be a domain with a Lipschitz boundary and let $u \in [W^{1,2}(\Omega)]^m$ be a solution to the system* (5.1.1). *Let $f^r \in L^2(\Omega)$ and let the conditions* (5.2.8)–(5.2.10) *be satisfied. Let σ be a function with the above mentioned properties. Then*

$$\int_\Omega \frac{\partial^2 u_r}{\partial x_i \, \partial x_j} \frac{\partial^2 u_r}{\partial x_i \, \partial x_j} \sigma^2 \, dx \leqq c[\|u\|^2_{[W^{1,2}(\Omega)]^m} + \|f\sigma\|_{[L^2(\Omega)]^m}]. \tag{5.2.12}$$

5.2.13. REMARK. It can be proved, see the work by J. Nečas [48], that for Ω with a Lipschitz boundary, there exists σ with the required properties and such that

$$c_1 \operatorname{dist}(x, \partial\Omega) \leqq \sigma(x) \leqq c_2 \operatorname{dist}(x, \partial\Omega), \tag{5.2.14}$$

$$|D^i\sigma(x)| \leqq c[\operatorname{dist}(x, \partial\Omega)]^{1-|i|}. \tag{5.2.15}$$

The proof of Theorem 5.2.11: Put $\tau e^l = \Delta x$. Let $|\tau| < h$, h fixed. If $v \in [W_0^{1,2}(\Omega)]^m$ with supp $v \subset \Omega$, then

$$\begin{aligned} \int_\Omega \{[a_i^r(x + \Delta x, u(x + \Delta x), \nabla u(x + \Delta x)) \\ - a_i^r(x, u(x), \nabla u(x))] \frac{\partial v_r}{\partial x_i}(x) \\ + [a^r(x + \Delta x, u(x + \Delta x), \nabla u(x + \Delta x)) \\ - a^r(x, u(x), \nabla u(x))] \, v_r(x) \\ - [f^r(x + \Delta x) - f^r(x)] \, v_r(x)\} \, dx = 0. \end{aligned} \tag{5.2.16}$$

Put $v_r(x) = \Delta_l^\tau u_r(x) \, \sigma_h^2(x)$. Put further $\dfrac{1}{\tau}[u(x + \Delta x) - u(x)] = \Delta u(x)$. For almost all x in Ω we have

$$\begin{aligned} & a_i^r(x + \Delta x, u(x + \Delta x), \nabla u(x + \Delta x)) - a_i^r(x, u(x), \nabla u(x)) \\ & = \int_0^1 \Big(\frac{\partial a_i^r}{\partial \eta_j^s}(x + t\,\Delta x, u(x) + t(u(x + \Delta x) - u(x)), \nabla u(x) \\ & \quad + t(\nabla u(x + \Delta x) - \nabla u(x)))\Big) \, dt \frac{\partial}{\partial x_j}[u_s(x + \Delta x) - u_s(x)] \\ & \quad + \left(\int_0^1 \frac{\partial a_i^r}{\partial \xi_s^r} \, dt\right) \cdot \tau \, \Delta u_s + \left(\int_0^1 \frac{\partial a_i^r}{\partial x_l} \, dt\right) \tau, \end{aligned} \tag{5.2.17}$$

where we do not write the arguments for the sake of simplicity. So we get

$$\int_\Omega \left(\int_0^1 \frac{\partial a_i^r}{\partial \eta_j^s} \, dt \right) \frac{\partial \Delta u_s}{\partial x_j} \frac{\partial \Delta u_r}{\partial x_i} \sigma_h^2 \, dx \tag{5.2.18}$$

$$= - \int_\Omega \left(\int_0^1 \frac{\partial a_i^r}{\partial \eta_j^s} \, dt \right) \frac{\partial \Delta u_s}{\partial x_j} 2\sigma_h \frac{\partial \sigma_h}{\partial x_i} \Delta u_r \, dx$$

$$- \int_\Omega \left(\int_0^1 \frac{\partial a_i^r}{\partial \xi_s} \, dt \right) \Delta u_s \left(\frac{\partial \Delta u_r}{\partial x_i} \sigma_h^2 + \Delta u_r \, 2\sigma_h \frac{\partial \sigma_h}{\partial x_i} \right) dx$$

$$- \int_\Omega \left(\int_0^1 \frac{\partial a_i^r}{\partial x_i^r} \, dt \right) \left(\frac{\partial \Delta u_r}{\partial x_i} \sigma_h^2 + \Delta u_r \, 2\sigma_h \frac{\partial \sigma_h}{\partial x_i} \right) dx$$

$$- \int_\Omega \left(\int_0^1 \frac{\partial a^r}{\partial \eta_j^s} \, dt \right) \frac{\partial \Delta u_s}{\partial x_j} \Delta u_r \sigma_h^2 \, dx - \int_\Omega \left(\int_0^1 \frac{\partial a^r}{\partial \xi_s} \, dt \right) \Delta u_s \Delta u_r \sigma_h^2 \, dx$$

$$- \int_\Omega \left(\int_0^1 \frac{\partial a^r}{\partial x_l} \, dt \right) \Delta u_r \sigma_h^2 \, dx + \int_\Omega \Delta f^r \Delta u_r \sigma_h^2 \, dx .$$

Denote the term on the left-hand side by J and the terms on the right-hand side by $J_1, J_2, \ldots, J_9$. We have

$$J_9 = - \int_\Omega f^r \Delta^{-\tau}(\Delta^\tau u_r \sigma_h^2) \, dx , \tag{5.2.19}$$

in virtue of Lemma 5.2.1 we thus get

$$|J_9| \leqq c_9 \Big[\Big(\int_\Omega f^r(x) f^r(x) \, \sigma_h^2(x - \Delta x) \, dx \Big)^{\frac{1}{2}} + \|f\sigma_h\|_0^2 + \|u\|_1^2 \Big] \stackrel{\text{def}}{=} c_9 [\|f\sigma_h^\tau\|_0^2 + \|f\sigma_h\|_0^2 + \|u\|_1^2] \tag{5.2.20}$$

(we write $\|u\|_1 = \|u\|_{[W^{1,2}(\Omega)]^m}$ for brevity).

Now we have

$$J \geqq c_0 \int_\Omega \frac{\partial \Delta u_r}{\partial x_i} \frac{\partial \Delta u_s}{\partial x_j} \sigma_h^2 \, dx \stackrel{\text{def}}{=} c_0 J^\tau , \tag{5.2.21}$$

$$\begin{aligned} &|J_1| \leqq c_1 J_\tau^{\frac{1}{2}} \|u\|_1 , && |J_2| \leqq c_2 \|u\|_1 J_\tau^{\frac{1}{2}} , \\ &|J_3| \leqq c_3 \|u\|_1^2 , && |J_4| \leqq c_4 \|u\|_1 J_\tau^{\frac{1}{2}} , \\ &|J_5| \leqq c_5 \|u\|_1^2 , && |J_6| \leqq c_6 J_\tau^{\frac{1}{2}} \|u\|_1 , \\ &|J_7| \leqq c_7 \|u\|_1^2 , && |J_8| \leqq c_8 \|u\|_1^2 , \end{aligned} \tag{5.2.22}$$

hence (5.1.18)–(5.1.22) imply

$$J_\tau \leqq c_{10} [\|f\sigma_h\|_0^2 + \|f\sigma_h^\tau\|_0^2 + \|u\|_1^2] . \tag{5.2.23}$$

Now

$$\left\| \Delta \left(\frac{\partial u}{\partial x_i} \sigma_h \right) \right\|_0 \leqq c_{11}(J_\tau^{\frac{1}{2}} + \|u\|_1), \tag{5.2.24}$$

hence (5.2.12) for σ_h follows from (5.2.23), (5.2.24) and from Lemma 5.2.4. We get

$$\int_\Omega \frac{\partial^2 u_r}{\partial x_i \, \partial x_j} \frac{\partial^2 u_r}{\partial x_i \, \partial x_j} \sigma_h^2 \, \mathrm{d}x \leqq c_{12}[\|u\|_1^2 + \|f\sigma\|^2] \tag{5.2.25}$$

and Fatou's lemma gives the result as $h \to 0$.

Let us now consider a cylindrical domain

$$C = \{x \in \mathbb{R}^n \mid x' = (x_1, \ldots, x_{n-1}), |x'| < 1, 0 < x_n < 1\}$$

and let $\Gamma = \{x \in \partial C \mid x_n = 0, |x'| < 1\}$. Let a matrix a_{lk} be given (see 3.1), where the subscript l runs through the set of subscripts $0 \leqq l_1 < l_2 < \ldots < l_{m_1} \leqq m$, the subscript k runs through the set $0 \leqq k_1 < k_2 < \ldots < k_{m_2} \leqq m$, and these two sets of subscripts are complementary. Suppose that $a_{lk} \in C^2(\bar{\Gamma})$ and let $V = [v \in \{W^{1,2}(C)]^m \mid v = 0$ on $\partial C \setminus \Gamma$, $v_l = a_{lk} v_k$ on $\Gamma\}$. Let functions $A^r(x, \xi)$ be given, that are continuously differentiable on $\bar{\Gamma} \times \mathbb{R}^m$ and such that

$$\sum_{r,l} \left| \frac{\partial A^r}{\partial x_l} (x, \xi) \right| + \sum_r |A^r(x, \xi)| \leqq c|\xi|, \tag{5.2.26}$$

$$\sum_{r,s} \left| \frac{\partial A^r}{\partial \xi_s} (x, \xi) \right| \leqq c. \tag{5.2.27}$$

Let further a function $u^0 \in [W^{2,2}(C)]^m$ be given and $g \in [W^{1,2}(C)]^m$.

Let us consider now $u \in [W^{1,2}(C)]^m$ such that

$$u - u^0 \in V, \tag{5.2.28}$$

$$\forall v \in V: \int_C \left[a_i^r(x, u, \nabla u) \frac{\partial v_r}{\partial x_i} + a^r(x, u, \nabla u) - f^r v_r \right] \mathrm{d}x + \int_\Gamma [A^r(x, u) - g^r] \, v_r \, \mathrm{d}x' = 0. \tag{5.2.29}$$

Let us consider $\sigma \in W^{1,\infty}(C)$, $\sigma = 0$ on $\partial C \setminus \Gamma$, $\sigma \geqq 0$, and let us suppose that there exists $\sigma_h \in C^\infty(\bar{C})$, such that $\sigma_h \leqq \sigma$ in C, $\|\sigma_h\|_{W^{1,\infty}(C)} \leqq c$, $\operatorname{supp} \sigma_h \subset \{x \in C \mid \operatorname{dist}(x, \partial C \setminus \Gamma) > h\} \overset{\text{def}}{=} C^h$, $(h > 0)$, $\sigma^h \to \sigma$ in $C(\bar{C})$.

5.2.30. THEOREM. *Let C be a cylindrical domain introduced as above and u be the weak solution to* (5.2.28), (5.2.29). *Let the conditions* (5.2.8)–(5.2.10), (5.2.26), (5.2.27) *be satisfied. Then*

$$\int_C \frac{\partial^2 u_r}{\partial x_i \, \partial x_j} \frac{\partial^2 u_r}{\partial x_i \, \partial x_j} \sigma^2 \, \mathrm{d}x \leqq c[\|u\|^2_{[W^{1,2}(C)]^m} + \|f\sigma\|^2_{[L^2(C)]^m} + \|\nabla u^0 \sigma\|^2_{[W^{1,2}(C)]^{mn}} + \|g\sigma\|^2_{[W^{1,2}(C)]^m}]. \tag{5.2.31}$$

Proof. Let $v \in V$, supp $v \subset C^h$. Let $1 \leqq l \leqq n-1$, $\Delta x = e^l\tau$, $|\tau| < h$, and $|\tau|$ be small enough. Let us define, using the notation introduced above:

(5.2.32) $\quad w_k(x) = 0, \qquad w_l(x) = \tau\, \Delta a_{lk}[u_k(x+\Delta x) - u_k^0(x+\Delta x)]\,\sigma_h^2(x),$

(5.2.33) $\quad \omega_k(x) = 0,$

$$\omega_l(x) = [a_{lk}(x+\Delta x) - a_{hk}(x)]\,(u_h(x) - u_h^0(x))\,\sigma_h^2(x).$$

Put $v = \Delta(u - u^0)\cdot\sigma_h^2$; in general $v \notin V$. We have

(5.2.34) $$v_l(x) - w_l(x) = a_{lk}(x)\,[v_k(x) - w_k(x)]$$

so $v - w \in V$, and similarly

(5.2.35) $$v_l(x) - \omega_l(x) = a_{lk}(x+\Delta x)\,[v_k(x) - \omega_k(x)];$$

thus $v^\tau - \omega^\tau \in V$.

It follows from (5.2.29) that

(5.2.36)
$$\begin{aligned}
&\int_C [a_i^r(x+\Delta x, u(x+\Delta x), \nabla u(x+\Delta x)) \\
&\quad - a_i^r(x, u(x), \nabla u(x))\Big(\frac{\partial v_r}{\partial x_i}(x) - \frac{\partial \omega_r}{\partial x_i}(x)\Big)\,dx \\
&+ \int_C a_i^r(x, u(x), \nabla u(x))\Big(\frac{\partial w_r}{\partial x_i}(x) - \frac{\partial \omega_r}{\partial x_i}(x)\Big)\,dx \\
&+ \int_C [a^r(x+\Delta x, u(x+\Delta x), \nabla u(x+\Delta x)) \\
&\quad - a^r(x, u(x), \nabla u(x))]\,(v_r(x) - \omega_r(x))\,dx \\
&+ \int_C a^r(x, u(x), \nabla u(x))\,(w_r(x) - \omega_r(x))\,dx \\
&+ \int_\Gamma [A^r(x+\Delta x, u(x+\Delta x)) - A^r(x, u(x))]\,(v_r(x) - \omega_r(x))\,dx' \\
&+ \int_\Gamma A^r(x, u(x), \nabla u(x))\,(w_r(x) - \omega_r(x))\,dx' \\
&- \int_C [f^r(x+\Delta x) - f^r(x)]\,(v_r(x) - \omega_r(x))\,dx \\
&- \int_C f^r(x)\,(w_r(x) - \omega_r(x))\,dx \\
&- \int_\Gamma [g^r(x+\Delta x) - g^r(x)]\,(v_r(x) - \omega_r(x))\,dx' \\
&- \int_\Gamma g^r(x)\,(w_r(x) - \omega_r(x))\,dx' = 0.
\end{aligned}$$

The comparison of (5.2.16) with (5.2.36) leads first to the estimates of the new expressions in (5.2.36): (for simplicity we shall use the notation just introduced: $g^\tau(x) = g(x-\tau)$, and the notation from the proof of the previous theorem)

(5.2.37) $$\left|\frac{1}{|\tau|^2}\int_C [a_i^{r,-\tau} - a_i^r]\frac{\partial \omega_r}{\partial x_i}\,dx\right| \leqq c_1[J_\tau^{\frac{1}{2}} + \|u\|_1]\,[\|u\|_1 + \|u^0\|_1],$$

(5.2.38) $$\left|\frac{1}{|\tau|^2}\int_C a_i^r\Big(\frac{\partial w_r}{\partial x_i} - \frac{\partial \omega_r}{\partial x_i}\Big)\,dx\right| \leqq c_2\|u\|_1(\|u\|_1 + \|u^0\|_1 + J_\tau^{\frac{1}{2}} + J_\tau^{0\frac{1}{2}}),$$

where $J_\tau^0 \overset{\text{def}}{=} \int\limits_C \frac{\partial \Delta u_r^0}{\partial x_i} \frac{\partial \Delta u_s^0}{\partial x_j} \sigma_h^2 \, dx$,

(5.2.39) $$\left| \frac{1}{|\tau|^2} \int\limits_C [a^{r,-\tau} - a^r] \, \omega_r \, dx \right| \leqq c_3 [J_\tau^{\frac{1}{2}} + \|u\|_1] [\|u\|_1 + \|u^0\|_1],$$

(5.2.40) $$\left| \frac{1}{|\tau|^2} \int\limits_C a^r (w_r - \omega_r) \, dx \right| \leqq c_4 \|u\|_1 (\|u\|_1 + \|u^0\|_1).$$

For the estimates of the integrals over Γ we shall use the inequality

(5.2.41) $$\forall \varepsilon > 0, \quad \int\limits_\Gamma f^2 \, dx' \leqq \varepsilon \|f\|^2_{W^{1,2}(C)} + \lambda(\varepsilon) \, \|f\|^2_{L^2(C)};$$

this follows from ($f \in C^1(\bar{C})$)

$$f^2(x', 0) = f^2(x', \xi) - 2 \int\limits_0^\xi f(x', \eta) \frac{\partial f}{\partial x_n} (x', \eta) \, d\eta$$

$$\leqq f^2(x', \xi) + 2 \left(\int\limits_0^1 f^2(x', \eta) \, d\eta \right)^{\frac{1}{2}} \left(\int\limits_0^1 \frac{\partial f}{\partial x_n} (x', \eta)^2 \, d\eta \right)^{\frac{1}{2}},$$

thus

$$\int\limits_\Gamma f^2 \, dx' \leqq \int\limits_C f^2 \, dx + \frac{1}{\varepsilon} \int\limits_C f^2 \, dx + \varepsilon \int\limits_C \left(\frac{\partial f}{\partial x_n} \right)^2 dx.$$

In this manner we get

(5.2.42) $$\left| \frac{1}{\tau^2} \int\limits_\Gamma [A^{r,-\tau} - A^r] \, \omega_r \, dx' \right| \leqq c_5 [\|u\|_1 (\|u\|_1 + \|u^0\|_1) + \left(\int\limits_\Gamma \sigma_h^2 (\Delta u)^2 \, dx' \right)^{\frac{1}{2}} (\|u\|_1 + \|u^0\|_1)]$$
$$\leqq c_6 [\|u\|_1^2 + \|u^0\|_1^2 + \varepsilon J_\tau + \lambda_1(\varepsilon) \, \|u\|_1^2],$$

(5.2.43) $$\left| \frac{1}{\tau^2} \int\limits_\Gamma A^r (w_r - \omega_r) \, dx' \right|$$
$$\leqq c_7 \|u\|_1 \left[\left(\int\limits_\Gamma \sigma_h^2 (\Delta u)^2 \, dx' \right)^{\frac{1}{2}} + \left(\int\limits_\Gamma \sigma_h^2 (\Delta u^0)^2 \, dx' \right)^{\frac{1}{2}} \right]$$
$$\leqq c_8 \|u\|_1 [J_\tau^{\frac{1}{2}} + \|u\|_1 + J_\tau^{0\frac{1}{2}} + \|u^0\|_1].$$

It is furthermore clear that

(5.2.44) $$\left| \frac{1}{\tau^2} \int\limits_C (f^{r,-\tau} - f^r) \, \omega_r \, dx \right|$$
$$\leqq c_9 (\|f \sigma_h\|_0 + \|f \sigma_h^\tau\|) \, (\|u\|_1 + \|u^0\|_1),$$

(5.2.45) $$\left| \frac{1}{\tau^2} \int\limits_C f^r (w_r - \omega_r) \, dx \right| \leqq c_{10} \|f\|_0 \, (\|u\|_1 + \|u^0\|_1),$$

(5.2.46) $$\left|\frac{1}{\tau^2}\int_\Gamma g^r(w_r-\omega_r)\,dx'\right|$$
$$\leqq c_{11}\|g\sigma_h\|_1\left[\left(\int_\Gamma \sigma_h^2(\Delta u)^2\,dx'\right)^{\frac{1}{2}}+\left(\int_\Gamma \sigma_h^2(\Delta u^0)^2\,dx'\right)^{\frac{1}{2}}\right]$$
$$\leqq c_{12}\|g\sigma_h\|_1\,[J_\tau^{\frac{1}{2}}+\|u\|_1+J_\tau^{0\frac{1}{2}}+\|u^0\|_1],$$

(5.2.47) $$\left|\frac{1}{\tau^2}\int_\Gamma [g^{r,-\tau}-g^r]\,\omega_r\,dx'\right|$$
$$\leqq c_{13}(\|g\sigma_h\|_1+\|g\sigma_h^\tau\|_1)\left[\left(\int_\Gamma \sigma_h(\Delta u)^2\,dx'\right)^{\frac{1}{2}}\right.$$
$$\left.+\left(\int_\Gamma \sigma_h(\Delta u^0)^2\,dx'\right)^{\frac{1}{2}}\right]$$
$$\leqq c_{14}(\|g\sigma_h\|_1+\|g\sigma_h^\tau\|_1)\,[J_\tau^{\frac{1}{2}}+\|u\|_1+J_\tau^{0\frac{1}{2}}+\|u^0\|_1].$$

Moreover we have (we suppose without the loss of generality that $g^r(x', 1) = 0$)

(5.2.48) $$\int_\Gamma [g^{r,-\tau}-g^r]\,\omega_r\,dx' = -\int_C -\frac{\partial}{\partial x_n}[(g^{r,-\tau}-g^r)\,v_r]\,dx.$$

Now $\int_C\left[\frac{\partial g^{r,\tau}}{\partial x_n}-\frac{\partial g^r}{\partial x_n}\right]v_r\,dx = -\int_C \frac{\partial g^r}{\partial x_n}\Delta^{-\tau}((\Delta^\tau u_r-\Delta^\tau u_r^0)\,\sigma_h^2)\,dx$, thus

$$\left|\int_C \frac{\partial g^r}{\partial x_n}\Delta^{-\tau}(\Delta^\tau(u_r-u_r^0)\,\sigma_h^2)\,dx\right|$$
$$\leqq c_{15}(\|g\sigma_h\|_1+\|g\sigma_h^{-\tau}\|_1)\,(J_\tau^{\frac{1}{2}}+\|u\|_1+J_\tau^{0\frac{1}{2}}+\|u^0\|_1)$$

and

$$\left|\int_C \Delta g^r\frac{\partial v_r}{\partial x_n}\,dx\right| \leqq c_{16}\left(\int_C \sigma_h^2|\Delta g|^2\,dx\right)^{\frac{1}{2}}(J_\tau^{\frac{1}{2}}+\|u\|_1+J_\tau^{0\frac{1}{2}}+\|u^0\|_1).$$

The other terms in (5.2.36) are treated as in the proof of the previous theorem, and, at the same time, $v = \Delta u\sigma_h^2 + \Delta u^0\sigma_h^2$, therefore the terms $J_\tau^{0\frac{1}{2}}$ are to be taken into account. So we finally get the estimate

(5.2.49) $$J_\tau \leqq c_{17}\left[\|u\|_1^2+\|u^0\|_1^2+J_\tau^0+\|f\sigma_h\|_0^2+\|f\sigma_h^\tau\|_0^2\right.$$
$$\left.+\|g\sigma_h\|_1^2+\|g\sigma_h^\tau\|_1^2+\int_C \sigma_h^2|\Delta g|^2\,dx\right].$$

But $\lim_{\tau\to 0} J_\tau^0 = \int_C \frac{\partial^2 u_r^0}{\partial x_i\,\partial x_l}\frac{\partial^2 u_r^0}{\partial x_i\,\partial x_l}\sigma_h^2\,dx$ (no summation over l),

$\lim_{\tau\to 0}\int_C \sigma_h^2|\Delta g|^2\,dx = \int_C \sigma_h^2\left|\frac{\partial g}{\partial x_l}\right|^2 dx$ (we use the continuity in the mean) and clearly $\lim_{\tau\to 0}\|f\sigma_h^\tau\|_1 = \|f\sigma_h\|_1$, $\lim_{\tau\to 0}\|g\sigma_h^\tau\|_1 = \|g\sigma_h\|_1$, and thus we get for $l = 1, 2, \ldots, n-1$ that (no summation over l):

(5.2.50) $$\int_C \sigma^2\frac{\partial^2 u_r}{\partial x_i\,\partial x_l}\frac{\partial^2 u_r}{\partial x_i\,\partial x_l}\,dx$$
$$\leqq c[\|u\|_1^2+\|f\sigma\|_0^2+\|\nabla u^0\sigma\|_1+\|g\sigma|_1^2].$$

The preceding theorem implies that $u \in [W^{2,2}_{\mathrm{loc}}(\Omega)]^m$, hence from Green's theorem and (5.2.29) we obtain

$$(5.2.51) \qquad -\frac{\partial a_i^r}{\partial \eta_j^s}\frac{\partial^2 u_s^r}{\partial x_i\,\partial x_j} - \frac{\partial a_i^r}{\partial \xi_s}\frac{\partial u_s}{\partial x_i} - \frac{\partial a_i^r}{\partial x_i} + a^r = f^r, \qquad r = 1, 2, \ldots, m,$$

which we can write as

$$(5.2.52) \qquad -\frac{\partial a_i^r}{\partial \eta_j^s}\frac{\partial^2 u_s}{\partial x_i\,\partial x_j} = F^r$$

with

$$(5.2.53) \qquad \|F\sigma\|_0 \leqq c[\|f\sigma\|_0 + \|u\|_1].$$

Now we can calculate $\dfrac{\partial^2 u_s}{\partial x_n^2}$, $s = 1, 2, \ldots, m$, from (5.2.52) and estimate the corresponding $L^2(C)$ norms.

Let us now consider a bounded domain Ω with a Lipschitz boundary, let $\Gamma \subset \partial\Omega$, and, in agreement with the definition of the Lipschitz boundary, let $\Gamma = \{x \in \mathbb{R}^n \mid x = (x', x_n),\ x_n = a(x'),\ |x'| < \alpha\}$. Let us suppose that $O \overset{\text{def}}{=} \{x \in \mathbb{R}^n \mid |x'| < \alpha,\ a(x') < x_r < a(x') + \beta\} \subset \Omega$. So let us consider the problem (5.2.28), (5.2.29) on Ω anew, provided that on O all the conditions on coefficients and data as in Theorem 5.2.30 are fulfilled. We easily get:

5.2.54. THEOREM. *Let $O \subset \Omega$ as was described above, and let $a \in C^2(\overline{B_\alpha(O)})$. Let the conditions* (5.2.8)–(5.2.10) *be satisfied in O and the conditions* (5.2.26), (5.2.27) *on Γ. Let again $V = \{v \in [W^{1,2}(\Omega)]^m \mid v = 0$ on $\partial O \setminus \Gamma$, $v_l = a_{lk}v_k$ on $\Gamma\}$. Suppose that $a_{lk} \in C^2(\bar{\Gamma})$. (In the sense that $a_{lk}(x', a(x')) \in C^{(2)}(\overline{B_\alpha(O)})$.) Let $f \in [L^2(O)]^m$, $g \in [W^{1,2}(O)]^m$, $u^0 \in W^{2,2}(O)]^m$. Let $\sigma \in W^{1,\infty}(O)$, $\sigma = 0$ on $\partial O \setminus \Gamma$, $\sigma \geqq 0$ and let for $h > 0$ there exist $\sigma_h \in C^\infty(\bar{O})$ such that $\sigma_h \leqq \sigma$, $\|\sigma_h\|_{W^{1,\infty}(0)} \leqq c$, $\operatorname{supp}\sigma_h \subset \{x \in O \mid \operatorname{dist}(x, \partial O \setminus \Gamma) > h\}$, $\sigma_h \to \sigma$ in $C(\bar{O})$. Then*

$$(5.2.55) \qquad \int_O \sigma^2 \frac{\partial^2 u_r}{\partial x_i\,\partial x_j}\frac{\partial^2 u_r}{\partial x_i\,\partial x_j}\,\mathrm{d}x \leqq [\|u\|_{[W^{1,2}(O)]^m} + \|u^0\|_{[W^{1,2}(O)]^m}$$
$$+ \|f\sigma\|^2_{[L^2(\Omega)]^m} + \|\nabla u^0\sigma\|^2_{[W^{1,2}(O)]^{mn}} + \|g\sigma\|^2_{[W^{1,2}(O)]^m}].$$

Proof. Let $y = y(x)$ be the diffeomorphism of $\bar{O}$ onto $\bar{C} = \{y = (y', y_n) \mid |y'| < \alpha, 0 < y_n < \beta\}$ defined by the relations

$$(5.2.56) \qquad y' = x', \qquad y_n = x_n - a(x').$$

If $u \in [W^{1,2}(O)]^m$, let $\tilde{u}(y) \overset{\text{def}}{=} u(x(y))$ and so on. We have $\dfrac{\mathrm{d}x}{\mathrm{d}y} = 1$ (the Jacobian), $\dfrac{\mathrm{d}S}{\mathrm{d}y'} = \sqrt{1 + \sum_{i=1}^{n-1}\left(\dfrac{\partial a}{\partial x_i}\right)^2}(y')$, and let us denote the matrix $\dfrac{\partial y_i}{\partial x_j}$ by $\dfrac{\partial y}{\partial x}$. We get

$$(5.2.57) \qquad \tilde{a}_k^r(y, \tilde{u}, \nabla_y\tilde{u}) = a_i^r\left(x, \tilde{u}, \nabla_y\tilde{u}\frac{\partial y}{\partial x}\right)\frac{\partial y_k}{\partial x_i}\left|\frac{\partial x}{\partial y}\right|,$$

$$(5.2.58) \qquad \tilde{a}^r(y, \tilde{u}, \nabla_y\tilde{u}) = a^r\left(x, \tilde{u}, \nabla_y\tilde{u}\frac{\partial y}{\partial x}\right),$$

(5.2.59) $$\tilde{f}^r(y) = f^r(x),$$

(5.2.60) $$\tilde{a}_{lk}(y) = a_{lk}(y', a(y')),$$

(5.2.61) $$\tilde{g}^r(y) = g^r(x)\frac{dS}{dy'},$$

(5.2.62) $$\tilde{A}^r(y, \tilde{u}) = A^r(x, \tilde{u})\frac{dS}{dy'},$$

(5.2.63) $$\tilde{V} = \{\tilde{v} \in [W^{1,2}(C)]^m \mid v = 0 \text{ on } \partial C \setminus y(\Gamma),\ \tilde{v}_l = \tilde{a}_{lk}\tilde{v}_k \text{ on } y(\Gamma)\}.$$

Then we have

(5.2.64) $$\tilde{u} - \tilde{u}^0 \in V,$$

(5.2.65) $$\forall \tilde{v} \in \tilde{V} \quad \text{it is} \int_C \left(\tilde{a}_i^r \frac{\partial \tilde{v}_r}{\partial y_i} + \tilde{a}^r\tilde{v}_r - \tilde{f}^r\tilde{v}_r\right) dy + \int_{y(\Gamma)} [\tilde{A}^r - \tilde{g}^r]\, \tilde{v}_r\, dy' = 0.$$

If we put $\tilde{\sigma}(y) = \sigma(x)$, then all the conditions of Theorem 5.2.30 are satisfied; for example, $\dfrac{\partial \tilde{a}_k^r}{\partial \tilde{\eta}_l^s} = \dfrac{\partial a_i^r}{\partial \eta_j^s}\dfrac{\partial y_k}{\partial x_i}\dfrac{\partial y_l}{\partial x_j}$, hence $\dfrac{\partial \tilde{a}_k^r}{\partial \tilde{\eta}_j^s}\tilde{\zeta}_k^r\tilde{\zeta}_l^s \geqq c_1|\tilde{J}|^2$.

Let us turn to the p-polynomial growth of the coefficients a_i^r, a^r, A^r. Let us mention, besides the books [1], [2], [33] just quoted, the paper by JU. A. DUBINSKIĬ [49] and the papers by J. P. GOSSEZ [35] and by M. MÜLLER, J. NEČAS [50].

Let the coefficients a_i^r, a^r be once continuously differentiable with respect to all the arguments in $\bar{\Omega} \times \mathbb{R}^m \times \mathbb{R}^{mn}$ and let

(5.2.66) $$\sum_{r,i,l}\left|\frac{\partial a_i^r}{\partial x_l}\right| + \sum_{r,i}|a_i^r| + \sum_r |a^r| + \sum_{r,l}\left|\frac{\partial a^r}{\partial x_l}\right| \leqq c(1 + \alpha|\xi| + |\eta|)^{p-1},$$

(5.2.67) $$\sum_{r,i,s,j}\left|\frac{\partial a_i^r}{\partial \eta_j^s}\right| + \sum_{r,i,s}\left|\frac{\partial a_i^r}{\partial \xi_s}\right| + \sum_{r,s,j}\left|\frac{\partial a^r}{\partial \eta_j^s}\right| + \sum_{r,s}\left|\frac{\partial a^r}{\partial \xi_s}\right| \leqq c(1 + \alpha|\xi| + |\eta|)^{p-2},$$

(5.2.68) $$\frac{\partial a_i^r}{\partial \eta_j^s}(x, \xi, \eta)\, \zeta_i^r\zeta_j^s \geqq c(1 + \alpha|\xi| + |\eta|)^{p-2}|\zeta|^2,$$

$\alpha \geqq 0$; we can also consider the right-hand side $\geqq c(\alpha|\xi| + |\eta|)^{p-2}\,|\zeta|^2$ in the last inequality, see the papers quoted above and many others.

5.2.69. THEOREM. *Let Ω be a domain with a Lipschitz boundary and let $u \in [W^{1,p}(\Omega)]^m$, $p \geqq 2$, be a weak solution to the system*

(5.2.70) $$-\frac{\partial}{\partial x_i}[a_i^r(x, u, \nabla u)] + a^r(x, u, \nabla u) = f^r, \qquad r = 1, 2, \ldots, m.$$

Suppose that the conditions (5.2.66)–(5.2.68) *are fulfilled. Let $f \in [L^2(\Omega)]^m$ and σ be such as in Theorem* 5.2.11. *Then*

(5.2.71) $$\int_\Omega \frac{\partial^2 u_r}{\partial x_i \partial x_j}\frac{\partial^2 u_r}{\partial x_i \partial x_j}(1 + \alpha|u|^{p-2} + |\nabla u|^{p-2})\,\sigma^2\, dx$$
$$\leqq c[\|u\|^2_{[W^{1,p}(\Omega)]^m} + \|f\sigma\|^2_{[L^2(\Omega)]^m}].$$

The proof is similar to the proof of Theorem 5.2.11, so we shall be more concise. If we write $W(\xi, \eta) = 1 + \alpha|\xi| + |\eta|$, we obtain for the integrals $J, J_1, \ldots, J_9$ from (5.2.18) that

(5.2.72) $$J \geqq \int_\Omega \int_0^1 W(u + t(u^{-\tau} - u), \nabla u + t(\nabla u^{-\tau} - \nabla u))^{p-2}\,dt \times \frac{\partial \Delta u_r}{\partial x_i}\frac{\partial \Delta u_r}{\partial x_i}\sigma_h^2\,dx \overset{\text{def}}{=} J_\tau,$$

(5.2.73) $$|J_1| \leqq c_1 J_\tau^{\frac{1}{2}}\left(\|u\|_{1,p} + \|u\|_{1,p}^{\frac{p}{2}}\right),$$

(5.2.74) $$|J_2| \leqq c_2 J_\tau^{\frac{1}{2}}\left(\|u\|_{1,p} + \|u\|_{1,p}^{\frac{p}{2}}\right),$$

(5.2.75) $$|J_3| \leqq c_3\left(\|u\|_{1,p} + \|u\|_{1,p}^{\frac{p}{2}}\right),$$

(5.2.76) $$|J_4| \leqq c_4 J_\tau^{\frac{1}{2}}\left(1 + \|u\|_{1,p}^{\frac{p}{2}}\right),$$

(5.2.77) $$|J_5| \leqq c_5(\|u\|_{1,p} + \|u\|_{1,p}^p),$$

(5.2.78) $$|J_6| \leqq c_6 J_\tau^{\frac{1}{2}}\left(\|u\|_{1,p} + \|u\|_{1,p}^{\frac{p}{2}}\right),$$

(5.2.79) $$|J_7| \leqq c_7(\|u\|_{1,p}^2 + \|u\|_{1,p}^p),$$

(5.2.80) $$|J_8| \leqq c_8(\|u\|_{1,p} + \|u\|_{1,p}^p),$$

(5.2.81) $$|J_9| \leqq c_9\left(\|f\sigma_h\|_{0,2} J_\tau^{\frac{1}{2}} + \|f\sigma_h^\tau\|_{0,2}\,\|u\|_{1,p}\right).$$

Hence (5.2.72)–(5.2.81) imply

(5.2.81) $$J_\tau \leqq c_{10}[1 + \|u\|_{1,p}^p + \|u\|_{1,p}^2 + \|u\|_{1,p} + \|f\sigma_h\|_{0,2}^2 + \|f\sigma_h^\tau\|_{0,2}^2].$$

Now

(5.2.83) $$\left\|\Delta\left(\frac{\partial u}{\partial x_i}\sigma_h\right)\right\|_{0,2} \leqq c_{11}\left(J_\tau^{\frac{1}{2}} + \|u\|_{1,p}\right),$$

and we thus get

(5.2.84) $$\int_\Omega \sigma_h^2 \frac{\partial^2 u_r}{\partial x_i\,\partial x_j}\frac{\partial^2 u_r}{\partial x_i\,\partial x_j}\,dx \leqq c_{12}[1 + \|u\|_{1,p}^p + \|u\|_{1,p}^2 + \|u\|_{1,p} + \|f\sigma_h\|_{0,2}^2 + \|f\sigma_h^\tau\|_{0,2}^2].$$

Now we pass to the limit as $\tau_k \to 0$, thus $\sigma_h \Delta^{\tau_k} u \to \sigma_h \dfrac{\partial u}{\partial x_l}$ in $[W^{1,2}(\Omega)]^m$ and also $\dfrac{\partial}{\partial x_i}[\sigma_h \Delta^{\tau_k} u] \to \dfrac{\partial}{\partial x_i}\left[\sigma_h \dfrac{\partial u}{\partial x_l}\right]$ almost everywhere in Ω. But $\int_0^1 W(u + t(u^{-\tau} - u), \nabla u + t(\nabla u^{-\tau} - \nabla u))\,dt \to W^{p-2}(u, \nabla u)$ almost everywhere in Ω (for τ_k well chosen). From (5.2.82) and Fatou's lemma we get (5.2.71) for σ_h and, finally, the limit passage $h \to 0$ and once more Fatou's lemma give the result.

5.3. Estimates in the space $W^{1,\infty}$

We shall again consider the system

$$(5.3.1)\qquad -\frac{\partial}{\partial x_i}[a_i^r(x, u, \nabla u)] + a^r(x, u, \nabla u) = f^r, \quad r = 1, 2, \ldots, m,$$

and we shall, as usual, suppose the continuous differentiability of the coefficients a_i^r, a^r, with respect to all variables and the conditions (5.2.8)–(5.2.10). Furthermore we shall suppose a condition of asymptoticity in the following form:

$$(5.3.2)\qquad \frac{\partial a_i^r}{\partial \eta_j^s}(x, \xi, \eta) \overset{\text{def}}{=} a_{ij}^{rs}(x, \xi, \eta) = d_{ij}^{rs}(x, \xi, \eta) + z_{ij}^{rs}(x, \xi, \eta),$$

where $z_{ij}^{rs}(x, \xi, \eta) = 0$ for $|\eta| \geqq \varkappa > 0$ and $d_{ij}^{\varrho s}(x, \xi, \eta) = 0$ for $s \neq \varrho$ and $|\eta| \geqq \varkappa$, where ϱ is a fixed index. We shall follow the work by J. NEČAS and J. STARÁ [51]. For the sake of simplicity, we shall confine ourselves to dimensions $2 \leqq n \leqq 3$. The general case can be solved by a method similar to the proof of Theorem 5.4.9, see also [1].

5.3.3. THEOREM. *Let $u \in [W^{1,2}(\Omega)]^m$ be a weak solution to the equation* (5.3.1). *Let the conditions* (5.3.2), (5.2.8)–(5.2.10) *be satisfied. For $n = 3$, let $\mu > 0$ and put $p = \dfrac{n(2+\mu)}{n+\mu}$, $q = \dfrac{n(2+\mu)}{n-2}$; for $n = 2$ let $0 < \mu < \infty$, $1 < q < \infty$ be arbitrary. Then for $\Omega' \subset \bar{\Omega}' \subset \Omega$*

$$(5.3.4)\qquad \|\nabla u_\varrho\|_{[L^q(\Omega')]^{mn}} \leqq \varkappa|\Omega|^{\frac{1}{q}} + c(\Omega')(1+\mu)$$
$$\times\left(\|f^\varrho\|_{L^p(\Omega)} + \|f\|_{[L^2(\Omega)]^m} + \|u\|_{[W^{1,2}(\Omega)]^m}\right).$$

(Without the loss of generality, we suppose that $\partial\Omega$ is smooth enough.)

Proof. It is $p \geqq 2$, hence Theorem 5.2.11 implies $u \in [W^{2,2}_{\text{loc}}(\Omega)]^m$ and the estimate (5.2.12). Let $u' = \dfrac{\partial u}{\partial x_l}$, $l = 1, 2, \ldots, m$, be fixed. For ϕ from $[W_0^{1,2}(\Omega)]^m$ such that $\operatorname{supp}\phi \subset \Omega$ we get

$$(5.3.5)\qquad \int_\Omega \frac{\partial a_i^r}{\partial \eta_j^s}\frac{\partial \phi_r}{\partial x_i}\frac{\partial u_s'}{\partial x_j}\,dx = -\int_\Omega f^r\phi_r\,dx - \int_\Omega \frac{\partial a_i^r}{\partial \xi_s}\frac{\partial \phi_r}{\partial x_i}u_s'\,dx$$
$$-\int_\Omega \frac{\partial a_i^r}{\partial x_l}\frac{\partial \phi_r}{\partial x_i}\,dx - \int_\Omega a^r\phi_r\,dx.$$

Let $w \in [W_0^{1,2}(\Omega)]^m$ be the solution to the problem $-\Delta w_r = -a^r(x, u, \nabla u)$. We get by Theorem 5.2.30 that

$$(5.3.6)\qquad \|w\|_{[W^{2,2}(\Omega)]^m} \leqq c_0\|u\|_{[W^{1,2}(\Omega)]^m}.$$

For $n = 3$, it is $u \in [W^{1,t_1}_{\text{loc}}(\Omega)]^m$ with $\dfrac{1}{t_1} = \dfrac{1}{2} - \dfrac{1}{n}$; moreover, it is $p < t_1$. For $n = 2$, it is $u \in [W^{1,p}_{\text{loc}}(\Omega)]^m$. We consider the case of $n = 3$ in the sequel.

The easier case of $n = 2$ is left to the reader. So we have

$$\int_\Omega \frac{\partial a_i^r}{\partial \eta_j^s} \frac{\partial \phi_r}{\partial x_i} \frac{\partial u_s'}{\partial x_j} \, dx = \int_\Omega g_i^r \frac{\partial \phi_r}{\partial x_i} \, dx, \tag{5.3.7}$$

where for $\bar{\Omega}' \subset \Omega'' \subset \bar{\Omega}'' \subset \Omega$ it is

$$\|g\|_{[L^p(\Omega'')]^m} \leqq c_1[\|f^p\|_{L^p(\Omega)} + \|f\|^2_{[L^2(\Omega)]^m} + \|u\|_{[W^{1,2}(\Omega)]^m}]. \tag{5.3.8}$$

Let now $1 \geqq \eta \geqq 0$, $\eta \in \mathscr{D}(\Omega)$, $\eta = 1$ on Ω''', $\bar{\Omega}' \subset \Omega''' \subset \bar{\Omega}''' \subset \Omega''$ and put $\phi = \eta\psi$, $\psi \in [W_0^{1,2}(\Omega)]^m$. It follows from (5.3.7) for $v = \eta u'$ that

$$\int_\Omega \frac{\partial a_i^r}{\partial \eta_j^s} \frac{\partial \psi_r}{\partial x_i} \frac{\partial v_s}{\partial x_j} \, dx = \int_\Omega \frac{\partial a_i^r}{\partial \eta_j^s} \frac{\partial \psi_r}{\partial x_i} u_s' \frac{\partial \eta}{\partial x_j} \, dx \tag{5.3.9}$$

$$- \int_\Omega \frac{\partial a_i^r}{\partial \eta_j^s} \psi_r \frac{\partial \eta}{\partial x_i} \frac{\partial u_s'}{\partial x_j} \, dx + \int_\Omega g_i^r \psi_r \frac{\partial \eta}{\partial x_i} \, dx + \int_\Omega g_i^r \frac{\partial \psi_r}{\partial x_i} \eta \, dx.$$

Let us first consider the terms

$$- \int_\Omega \frac{\partial a_i^r}{\partial \eta_j^s} \frac{\partial \eta}{\partial x_i} \frac{\partial u_s'}{\partial x_j} \psi_r \, dx + \int_\Omega g_i^r \psi_r \frac{\partial \eta}{\partial x_i} \, dx = \int_\Omega h_r \psi_r \, dx,$$

where

$$\|h\|_{[L^2(\Omega)]^m} \leqq c_2[\|f^\varrho\|_{L^p(\Omega)} + \|u\|_{[W^{1,2}(\Omega)]^m}]. \tag{5.3.10}$$

For every r, let $\omega_r \in W_0^{1,2}(\Omega)$ be such that $-\Delta\omega_r = h_r$. Theorem 5.2.54 gives

$$\|\omega\|_{[W^{2,2}(\Omega)]^m} \leqq c_3 \|h\|_{[L^2(\Omega)]^m}. \tag{5.3.11}$$

Hence it follows from (5.3.9) and (5.3.11) that

$$\int_\Omega \frac{\partial a_i^r}{\partial \eta_j^s} \frac{\partial \psi_r}{\partial x_i} \frac{\partial v_s}{\partial x_j} \, dx = \int_\Omega k_i^r \frac{\partial \psi_r}{\partial x_i} \, dx \tag{5.3.12}$$

and

$$\|k\|_{[L^p(\Omega)]^{mn}} \leqq c_4(\|f^p\|_{L^p(\Omega)} + \|f\|_{[L^2(\Omega)]^m} + \|u\|_{[W^{1,2}(\Omega)]^m}). \tag{5.3.13}$$

Put now $\psi_r = 0$ for $r \neq \varrho$, $\psi_\varrho = \operatorname{sign} v_\varrho \min(L, \max(|v_\varrho|, \varkappa) - \varkappa)^{1+\mu} \overset{\text{def}}{=} h_\varrho^{1+\mu}$, where $0 < L < \infty$. It is $\psi \in [W_0^{1,2}(\Omega)]^m$ and if $|v_\varrho| \geqq \varkappa$, then $|u_\varrho'| \geqq \varkappa$, thus $|\eta| \geqq \varkappa$ and we can use the condition (5.3.2). It thus follows from (5.3.12) (summation over 3 indices!) that

$$\int_\Omega d_{ij}^{rs} h_r^\mu \frac{\partial h_r}{\partial x_i} \frac{\partial h_s}{\partial x_j} \, dx = \int_\Omega k_i^r h_r^\mu \frac{\partial h_r}{\partial x_i} \, dx \tag{5.3.14}$$

$$= \frac{1}{\left(1 + \frac{\mu}{2}\right)^2} \int_\Omega d_{ij}^{rs} \frac{\partial h_r^{1+\frac{\mu}{2}}}{\partial x_i} \frac{\partial h_s^{1+\frac{\mu}{2}}}{\partial x_j} \, dx \geqq \frac{c_5}{\left(1 + \frac{\mu}{2}\right)^2} \left\|\nabla h_\varrho^{1+\frac{\mu}{2}}\right\|^2_{[L^2(\Omega)]^n}.$$

Now $\int_\Omega k_i^r h_r^\mu \frac{\partial h_r}{\partial x_i} \, dx = \frac{1}{1 + \frac{\mu}{2}} \int_\Omega k_i^r h^{\frac{\mu}{2}} \frac{\partial h_r^{1+\frac{\mu}{2}}}{\partial x_i} \, dx$, from which (with a clear notation used)

$$(5.3.15)\qquad \left|\int_\Omega k_i^r h_r^\mu \frac{\partial h_r}{\partial x_i}\,\mathrm{d}x\right|_{q\frac{(n-2)\mu}{2n(2+\mu)}} \leqq \frac{c_5}{1+\frac{\mu}{2}}\|k\|_{0,p}\,\|h_\varrho\|_{0,p}\left\|\nabla h_\varrho^{1+\frac{\mu}{2}}\right\|_{0,2}.$$

Finally we obtain from (5.3.14) and (5.3.15) that

$$(5.3.16)\qquad \left\|\nabla h_\varrho^{1+\frac{\mu}{2}}\right\|_{0,2} \leqq c_6(1+\mu)\,\|k\|_{0,p}\left\|h_\varrho\right\|_{0,q}^{q\frac{(n-2)\mu}{2n(2+\mu)}}.$$

Since $h_\varrho = 0$ on $\partial\Omega$, we have

$$(5.3.17)\qquad \left\|h_\varrho\right\|_{0,q}^{1+\frac{\mu}{2}} \leqq c_7\left\|\nabla h_\varrho^{1+\frac{\mu}{2}}\right\|_{0,2},$$

hence it follows from (5.3.13), (5.3.16), and (5.3.17) that

$$(5.3.18)\qquad \|h_\varrho\|_{0,q} \leqq c_8(1+\mu)\,\|k\|_{0,p}.$$

If we pass to the limit as $L\to\infty$ in (5.3.18), we get the result.

5.3.19. THEOREM. *Let $u\in[W^{1,2}(\Omega)]^m$ be a weak solution to the equation* (5.3.4), *let a_i^r, a^r satisfy the conditions* (5.2.8)–(5.2.10) *and* (5.3.2). *Let $f^\varrho\in L^p(\Omega)$, $p>n$. Let $\Omega'\subset\bar\Omega'\subset\Omega$. Then for $2\leqq n\leqq 3$ (see the remark in the beginning of this section) we have*

$$(5.3.20)\qquad \sup_{x\in\Omega'}|\nabla u_\varrho(x)| \leqq \varkappa + c(\Omega') \times\left(\|f\|_{[L^2(\Omega)]^m} + \|f^\varrho\|_{L^p(\Omega)} + \|u\|_{[W^{1,2}(\Omega)]^m}\right).$$

Proof. Considering the case of $n=3$, we choose (5.3.5) as our starting point. Without the loss of generality, we can suppose that $p<\dfrac{2n}{n-2}$. If $\bar\Omega'\subset\Omega''\subset\bar\Omega''\subset\Omega$, then $u\in\left[W^{1,\frac{2n}{n-2}}(\Omega'')\right]^m$ by virtue of Theorem 5.2.11 and the imbedding theorem, thus (5.3.7) and (5.3.8) are again true. The same argument leads to the equation (5.3.12) and to (5.3.13). Let h be defined as in the proof of the previous theorem. Let $\mu_0=0$, $\gamma=\dfrac{n(p-2)}{p(n-2)}$. $\mu_{k+1}=\gamma(2+\mu_k)$. From (5.3.12) we get, if we put $\psi_r=0$ for $r\neq\varrho$, $\psi_\varrho = h_\varrho^{1+\mu_k} \overset{\text{def}}{=} h^{1+\mu_k}$, $d_{ij}^{\varrho\varrho}=d_{ij}$ (no summation over ϱ), that

$$(5.3.21)\qquad \int_\Omega d_{ij}h^{\mu_k}\frac{\partial h}{\partial x_i}\frac{\partial h}{\partial x_j}\,\mathrm{d}x = \int_\Omega k_i^\varrho\frac{\partial h}{\partial x_i}h^\mu\,\mathrm{d}x.$$

Once more we have (5.3.14):

$$(5.3.21')\qquad \int_\Omega d_{ij}h^{\mu_k}\frac{\partial h}{\partial x_i}\frac{\partial h}{\partial x_j}\,\mathrm{d}x \geqq \frac{c_1}{\left(1+\frac{\mu_k}{2}\right)^2}\left\|\nabla h^{1+\frac{\mu_k}{2}}\right\|^2_{[L^2(\Omega)]^m}.$$

Now

$$\left|\int_\Omega k_i^\varrho \frac{\partial h}{\partial x_i} h^{\mu_k}\,\mathrm{d}x\right| \tag{5.3.22}$$

$$\leqq c_2\|k^\varrho\|_{L^p(\Omega)} \left\|\nabla h^{1+\frac{\mu_{k-1}}{2}}\right\|^\gamma_{[L^2(\Omega)]^m} \left\|\nabla h^{1+\frac{\mu_k}{2}}\right\|_{[L^2(\Omega)]^m},$$

where, because $h = 0$ on $\partial\Omega$, the imbedding theorem $W_0^{1,2}(\Omega) \subsetneq L^{\frac{2n}{n-2}}(\Omega)$ has been used. Hence it follows from (5.3.21) and (5.3.22) (with a clear notation employed) that

$$\left\|\nabla h^{1+\frac{\mu_k}{2}}\right\|_2 \leqq c_3\left(1+\frac{\mu_k}{2}\right)\|k^\varrho\|_p \left\|\nabla h^{1+\frac{\mu_{k-1}}{2}}\right\|_2^\gamma. \tag{5.3.23}$$

Let $\|k^\varrho\|_p \neq 0$. Put $z_k = \|k^\varrho\|_p^{-1}\left\|\Delta h^{1+\frac{\mu_k}{2}}\right\|_2^{\frac{1}{1+\frac{\mu_k}{2}}}$. Suppose that $c_3 \geqq 1$. It follows from (5.3.23) that

$$z_k = \left[c_3\left(1+\frac{\mu_k}{2}\right)\right]^{\frac{1}{1+\frac{\mu_k}{2}}} z_{k-1}^{\frac{\mu_k}{2+\mu_k}}. \tag{5.3.24}$$

Put $t_0 = 1 + z_0$ and

$$t_k = \left[c_3\left(1+\frac{\mu_k}{2}\right)\right]^{\frac{1}{1+\frac{\mu_k}{2}}} t_{k-1}. \tag{5.3.25}$$

Clearly $z_k \leqq t_k \leqq c_4 t_0$, which gives

$$\frac{\left\|\nabla h^{1+\frac{\mu_k}{2}}\right\|_2^{\frac{1}{1+\frac{\mu_k}{2}}}}{\|k^\varrho\|_p} \leqq c_4\left(1+\frac{\|\nabla h\|_2}{\|k^\varrho\|_p}\right). \tag{5.3.26}$$

Hence (5.3.26) and (5.3.17) imply

$$\|h\|_{q_k} \leqq c_5(\|k^\varrho\|_p + \|\nabla h\|_2), \tag{5.3.27}$$

where $q_k = \dfrac{n(2+\mu_k)}{n-2}$. Let $L \to \infty$ in (5.3.27). We get in such a way that

$$\left(\int_{\Omega'} [\max(|v_\varrho|, \varkappa) - \varkappa]^{q_k}\,\mathrm{d}x\right)^{\frac{1}{q_k}} \leqq c_5(\|k^\varrho\|_p + \|\nabla h\|_2); \tag{5.3.28}$$

the result follows as $k \to \infty$.

Let us turn to the estimate of $\|\nabla u\|_{[L^\infty(\Omega')]^m}$ when the case of $|\partial\Omega' \cap \partial\Omega| \neq 0$ can occur. For $m = 1$, such estimates are in the books [1] and [4]; here we shall again obtain such a result as a special case. Let C be again a cylinder: $C = \{x = (x', x_n) \in \mathbb{R}^n \mid |x'| < 1, 0 < x_n < 1\}$ and let $\Gamma = \{x \in \mathbb{R}^n \mid |x'| < 1, x_n = 0\}$. For the sake of simplicity, we suppose that the space V is defined as

$$V \overset{\text{def}}{=} \{v \in [W^{1,2}(C)]^m \mid v = 0 \text{ on } \partial C \setminus \Gamma,\ v_1 = v_2 = \ldots = v_{m_1} = 0 \text{ on } \Gamma,\ 0 \leqq m_1 \leqq m\}. \tag{5.3.29}$$

The general case of V given by the relations $v_l = a_{lk}v_k$ as above can be reduced to the previous one by the substitution $u_r = v_r$, where r runs through the set of subscripts $k_1, k_2, \ldots, k_{m_2}$, $u_l = v_l - a_{lk}v_k$, where l runs through the set of subscripts $l_1, l_2, \ldots, l_{m_1}$. We shall not discuss the generality of this point of view.

We first prove a theorem on tangential derivatives.

5.3.30. THEOREM. *Let C be a cylinder defined as above and V the space of test functions just defined. Let the coefficient a_i^r, a^r satisfy the conditions (5.2.8)–(5.2.10). Let the condition (5.3.2) be satisfied for $\varrho = 1, 2, \ldots, m$. Let further the functions $A^r(x, \xi)$ be once continuously differentiable with respect to all variables and let*

(5.3.31) $$\sum_{r=1}^{m} |A^r| + \sum_{r,i} \left| \frac{\partial A^r}{\partial x_i} \right| \leq c|\xi|,$$

(5.3.32) $$\sum_{r,s} \left| \frac{\partial A^r}{\partial \xi_s} \right| \leq c.$$

Let $f \in [L^p(C)]^m$, $g \in [W^{1,q}(\Gamma)]^{m-m_1}$, $p > n$, $q \geq 4$ for $n = 3$, $q > 1$ for $n = 2$. Let $u \in [W^{1,2}(C)]^m$ be a weak solution satisfying the relations

(5.3.33) $$u_r = u_r^0 \quad on\ \Gamma, \quad r = 1, 2, \ldots, m_1,$$

(5.3.34) $$\forall v \in V$$

$$\int_C \left[a_i^r \frac{\partial v_r}{\partial x_i} + a^r v_r - f^r v_r \right] \mathrm{d}x + \int_\Gamma [A^r - g^r]\, v_r \,\mathrm{d}x' = 0.$$

Let $2 \leq n \leq 3$ (see the above remark), $C' \subset C$, $\bar{C}' \subset C \cup \Gamma$. Then

(5.3.35) $$\sum_{i=1}^{n-1} \left\| \frac{\partial u}{\partial x_i} \right\|_{[L^\infty(C')]^m}$$

$$\leq C(C', \varkappa)\, [1 + \|u\|_{[W^{1,2}(C)]^m} + \|u^0\|_{[W^{2,p}(C)]^m} + \|f\|_{[L^p(C)]^m} + \|g\|_{[W^{1,q}(\Gamma)]^{m-m_1}}].$$

Proof. Consider $n = 3$. In virtue of Theorem 5.2.30, $u \in [W^{2,2}(C')]^m$ and

(5.3.36)

$$\|u\|_{[W^{2,2}(C')]^m} \leq c_1[\|u\|_{[W^{1,2}(C)]^m} + \|f\|_{[L^2(C)]^m} + \|u^0\|_{[W^{2,2}(\Omega)]^m} + \|g\|_{[W^{1,2}(C)]^m}],$$

where g is extended to the whole C by putting $g(x) = g(x')$ (for g_k we put $g_k \equiv 0$). Let $v' = \dfrac{\partial v}{\partial x_l}$, $l = 1, 2, \ldots, n-1$, $v \in [W^{2,2}(C)]^m$, and let $v \in V$, $\nabla v = 0$ on $\partial C \setminus \Gamma$. Then it follows from (5.3.34) as usual that

(5.3.37) $$\int_C \left[\frac{\partial a_i^r}{\partial \eta_j^s} \frac{\partial v_r}{\partial x_i} \frac{\partial u_s'}{\partial x_j} + \frac{\partial a_i^r}{\partial \xi_s} \frac{\partial v_r}{\partial x_i} u_s' + \frac{\partial a_i^r}{\partial x_l} \frac{\partial v_r}{\partial x_i} + \frac{\partial a^r}{\partial \eta_j^s} v_r \frac{\partial u_s'}{\partial x_j} \right.$$

$$\left. + \frac{\partial a^r}{\partial \xi_s} v_r u_s' + \frac{\partial a^r}{\partial x_l} v_r + f^r v_r' \right] \mathrm{d}x$$

$$+ \int_\Gamma \left[\frac{\partial A^r}{\partial \xi_s} v_r u_s' + \frac{\partial A^r}{\partial x_l} v_r - g^{r'} v_r \right] \mathrm{d}x' = 0.$$

The terms $\int_C$ are treated in (5.3.37) as in Theorem 5.3.19. Now Theorem 2.4.10 implies $u' \in [L^4(\Gamma)]^m$; suppose that $p < 6$ and $q < 4$. If $\bar{C}'' \cap \bar{\Gamma} \subset C \cup \Gamma$, then we obtain with the standard notation $\frac{\partial a_i^r}{\partial \eta_j^s} = a_{ij}^{rs}$, $\Gamma'' = \Gamma \cap \bar{C}''$, that

$$\int_{C''} a_{ij}^{rs} \frac{\partial v_r}{\partial x_i} \frac{\partial u'_s}{\partial x_j} \, dx = \int_C h_i^r \frac{\partial v_r}{\partial x_i} \, dx + \int_{\Gamma''} \lambda^r v_r \, dx', \tag{5.3.38}$$

where

$$\|h\|_{[L^p(C'')]^m} + \|\lambda\|_{[L^q(\Gamma'')]^{m-m_1}} \leqq c_1[\ ]; \tag{5.3.39}$$

we denote by [ ] the brackets on the right-hand side of (5.3.35). Let now $\eta \in \mathscr{D}(\mathbb{R}^n)$, $\eta = 1$ on $\bar{C}'$, and let $\operatorname{supp} \eta \cap \bar{C} \subset C'' \cup \Gamma''$. As in (5.3.9), put $u'\eta = w$, $v = \eta\psi$, $\psi \in V$, in (5.3.39). For the integrals $\int_C$ we have the same estimates as in the proof of Theorem 5.3.19 (we later come to the result), in the surface integral let us put only $\mu^r = \lambda^r \eta$, and we thus get

$$\int_{C''} a_{ij}^{rs} \frac{\partial \psi_r}{\partial x_i} \frac{\partial w_s}{\partial x_j} \, dx = \int_{C''} k_i^r \frac{\partial \psi_r}{\partial x_i} \, dx + \int_{\Gamma''} \mu^r \psi_r \, dx', \tag{5.3.40}$$

where

$$\|k\|_{[L^p(C'')]^{mn}} + \|\mu\|_{[L^q(\Gamma'')]^{m-m_1}} \leqq c_2[\ ]. \tag{5.3.41}$$

Put $\omega = w - u^{0\prime}\eta$. We finally get

$$\int_{C''} a_{ij}^{rs} \frac{\partial \psi_r}{\partial x_i} \frac{\partial \omega_s}{\partial x_j} \, dx = \int_{C''} v_i^r \frac{\partial \psi_r}{\partial x_i} \, dx + \int_{\Gamma''} \tilde{\mu}^r \psi_r \, dx', \tag{5.3.42}$$

$$\|v\|_{[L^p(C'')]^{mn}} + \|\tilde{\mu}\|_{[L^q(\Gamma'')]^{m-m_1}} \leqq c_3[\ ]. \tag{5.3.43}$$

Let $h_r \overset{\text{def}}{=} \min(L, \max(|\omega_r|, \varkappa) - \varkappa)$ and $\psi_r = \operatorname{sign} \omega_r h_r^{1+\mu_k}$. We proceed as in the proof of Theorem 5.3.19. The integrals $\int_{C''}$ are treated in the same manner. Using Theorem 2.4.10 once more, we obtain $\gamma = 3 - \frac{6}{p}$ and (with a clear notation used)

$$\left\| \nabla h^{1+\frac{\mu_k}{2}} \right\|_2 \leqq c_4[\|v\|_p + \|\tilde{\mu}\|_q] \left\| \nabla h^{1+\frac{\mu_{k-1}}{2}} \right\|_2^{\gamma} \left\| \nabla h^{1+\frac{\mu_k}{2}} \right\|_2. \tag{5.3.44}$$

We complete the proof as in Theorem 2.4.10.

Let us turn to the estimation of $\frac{\partial u}{\partial x_n}$. First, because $u \in [W^{2,2}_{\text{loc}}(C \cup \Gamma)]^m$, the condition (3.1.19) is realy satisfied and, in our special case, it has the form

$$-a_n^k(x, u, \nabla u) + A^k(x, u) = g^k, \quad k = m_1 + 1, \dots, m. \tag{5.3.45}$$

We shall suppose for the boundary conditions that it is possible to calculate $\frac{\partial u_r}{\partial x_n}$, $r = m_1 + 1, \dots, m$, from (5.3.45) by means of $\frac{\partial u_s}{\partial x_i}$, $s = 1, 2, \dots, m$, $i = 1, 2, \dots, n-1$ (so, roughly speaking, that no derivates $\frac{\partial u_r}{\partial x_n}$, $r = 1, 2, \dots, m_1$, occur in (5.3.45)) and that

(5.3.46)
$$\sum_{r=m_1+1}^{m} \left\| \frac{\partial u_r}{\partial x_n} \right\|_{L^\infty(\Gamma'')} \leqq c_5(\Gamma'') \cdot [\ ].$$

(The brackets from (5.3.35).)

Let us again consider (5.3.34), where, for the test function, we take $\dfrac{\partial v}{\partial x_n}$, $v \in [C^2(\bar{C})]^m$ such that supp $v \subset C \cup \Gamma$ and $\dfrac{\partial v_r}{\partial x_n} = 0$ on Γ for $r = 1, 2, \ldots, m_1$. So we get $\left(u' = \dfrac{\partial u}{\partial x_n}\right)$

(5.3.47)
$$\int_C \left[a_{ij}^{rs} \frac{\partial v_r}{\partial x_i} \frac{\partial u'_s}{\partial x_j} + \frac{\partial a_i^r}{\partial \xi_s} \frac{\partial v_r}{\partial x_i} u'_s + \frac{\partial a_i^r}{\partial x_n} \frac{\partial v_r}{\partial x_i} + \frac{\partial a^r}{\partial \eta_j^s} v_r \frac{\partial u'_s}{\partial x_j} \right.$$
$$\left. + \frac{\partial a^r}{\partial \xi_s} v_r u'_s + \frac{\partial a^r}{\partial x_n} v_r + f^r v'_r \right] \mathrm{d}x + \int_\Gamma \sum_{r=1}^{m_1} \sum_{i=1}^{n-1} - a_i^r \frac{\partial v_r}{\partial x_i} \mathrm{d}x' = 0,$$

where (5.3.45) is used. Let us actually consider only v such that $v_r = 0$, $r = m_1 + 1, \ldots, m$, that $\sum_{i=1}^{n-1} \dfrac{\partial}{\partial x_i} [a_i^r] \in L^q(\Gamma'')$, $r = 1, 2, \ldots, m_1$; and that

(5.3.48)
$$\sum_{r=1}^{m_1} \left\| \sum_{i=1}^{n-1} \frac{\partial}{\partial x_i} [a_i^r] \right\|_{L^q(\Gamma'')} \leqq c_6(\Gamma'') [\ ].$$

Let the Conditions (5.3.46) and (5.3.48): be called supplementary conditions.

For orientation, let us consider the Dirichlet and Neumann problem. In the first case we have to estimate $\sum_{r=1}^{m} \left\| \sum_{i=1}^{n-1} \dfrac{\partial}{\partial x_i} a_i^r \right\|_{L^q(\Gamma'')}$ which is restrictive enough. If $m = 1$, then it is possible to estimate $\dfrac{\partial a_i}{\partial \eta_j} \dfrac{\partial u}{\partial x_i} \nu_j$, see [1] or [4], using the maximum principle technique. In the case of the Neumann problem, our result is general, because from (5.3.45) we can calculate and estimate all $\dfrac{\partial u_r}{\partial x_n}$: the matrix $\dfrac{\partial a_n^k}{\partial \eta_n^l}$ is invertible and our conditions imply (5.3.46).

In general, as the reader sees, the $W^{1,\infty}$ estimates still include open problems, because both the asymptoticity conditions and the supplementary conditions are discussible.

5.3.49. THEOREM. *Let the conditions of Theorem* 5.3.32 *be fulfilled. Let the supplementary conditions* (5.3.46) *and* (5.3.48) *be satisfied. Then we have for* $C' \subset C \cup \Gamma$ *that*

(5.3.50)
$$\sum_{i=1}^{n} \left\| \frac{\partial u}{\partial x_i} \right\|_{[L^\infty(C')]^m}$$
$$\leqq c(C') [1 + \|u\|_{[W^{1,2}(C)]^m} + \|u^0\|_{[W^{2,p}(C)]^m} + \|f\|_{[L^p(C)]^m} + \|g\|_{[W^{1,q}(\Gamma)]^{m-m_1}}],$$
$$2 \leqq n \leqq 3.$$

Proof. First we have (5.3.35). Consider $n = 3$. Now we treat (5.3.47) in the same manner as (5.3.37). We thus come to (5.3.42) where $\psi_r = 0$ on Γ for $r = m_1 + 1, \ldots, m$. Now we define the functions h_r, $r = 1, 2, \ldots, m$, for $\varkappa' \geqq \varkappa$, where $\varkappa' \geqq \max\left(\varkappa, \left\|\frac{\partial u_{m_1+1}}{\partial x_n}\right\|_{L^\infty(\Gamma'')}, \ldots, \left\|\frac{\partial u_m}{\partial x_n}\right\|_{L^\infty(L'')}\right)$.

5.4. Maximum principle and the Liouville theorem, $m = 1$

The reader can omit this section if he is familiar with [1], [2], or [4] because there is nothing to much new here-only some simplifications. The purpose of this section is to accent the Liouville type theorem which is the basic means for the study of the regularity of systems. The real Liouville theorem follows from the Harnack inequality and we refer to the books [1], [4], [6], and to the works by J. MOSER [52], [53]. The simplicity of our exposition follows from the replacement of the Harnack theorem by its weak version as is done in [2], which we shall loosely follow.

We begin with really easy things:

consider $\Omega \subset \mathbb{R}^n$, $a_{ij} \in L^\infty(\Omega)$, such that

$$a_{ij}\xi_i\xi_j \geqq c|\xi|^2, \quad c > 0 \tag{5.4.1}$$

and consider weak solutions $w \in W^{1,2}(\Omega)$ such that

$$-\frac{\partial}{\partial x_i}\left[a_{ij}\frac{\partial w}{\partial x_j}\right] = 0. \tag{5.4.2}$$

5.4.3. DEFINITION. *A function $w \in W^{1,2}(\Omega)$ (Ω bounded) is called a weak subsolution (supersolution) to the equation* (5.4.2) *if*

$$\int_\Omega a_{ij}\frac{\partial \phi}{\partial x_i}\frac{\partial w}{\partial x_j}\,dx \leqq 0 \quad (\geqq 0) \tag{5.4.4}$$

$\forall \phi \in W_0^{1,2}(\Omega)$, $\phi \geqq 0$ *in* Ω.

5.4.5. THEOREM. *Let Ω be a domain with a Lipschitz boundary. Let w be a weak subsolution to the equation* (5.4.2) *and let* $\sup_{x\in\partial\Omega} u(x) = M < \infty$. *Then*

$$\sup_{x\in\Omega} u(x) \leqq \sup_{x\in\partial\Omega} u(x). \tag{5.4.6}$$

Proof. Let $\Omega_M = \{x \in \Omega \mid w(x) > M\}$. Put $\phi = \max(M, w(x)) - M$. It follows from (5.4.4) that

$$\int_\Omega a_{ij}\frac{\partial \phi}{\partial x_i}\frac{\partial \phi}{\partial x_j}\,dx = 0. \tag{5.4.7}$$

Thus ϕ is constant and because $\phi = 0$ on $\partial\Omega$, $\phi \equiv 0$ in Ω.

5.4.8. EXERCISE. For supersolutions, we have the minimum principle.

We prove a certain analogue to Theorem 5.3.19 and for $n \geqq 2$. The method can be used for a generalization of Theorem 5.3.19 to all dimensions.

5.4.9. THEOREM. *Let* $u \in W^{1,2}(B_{2R}(x^0))$ *be a weak non-negative subsolution to the equation* (5.4.2) *where* $a_{ij} \in L^\infty(\mathbb{R}^n)$ *and let the condition* (5.4.1) *be satisfied in* $\mathbb{R}^n$. *Then*

$$\sup_{x \in B_R(x^0)} u(x) \leqq c_1(2R)^{-\frac{n}{2}} \Big(\int_{B_{2R}(x^0)} u^2 \, dx\Big)^{\frac{1}{2}}. \tag{5.4.10}$$

Proof. Consider first $0 < \varrho \leqq 2R$, $\eta \in \mathscr{D}(B_\varrho(x^0))$; $\eta \geqq 0$. We have

$$\int_{B_\varrho} a_{ij} \frac{\partial \phi}{\partial x_i} \frac{\partial u}{\partial x_j} \, dx \leqq 0 \tag{5.4.11}$$

for $\forall \phi \in W_0^{1,2}(B_\varrho)$, $\phi \geqq 0$. Put

$$\phi = [\min(u(x), m)]^{1+\tau} \eta^2 \tag{5.4.12}$$

for $\tau \geqq 0$, $m > 0$. Writing $h = \min(u(x), m)$, we get from (5.4.11) that

$$c_1 \Big(\int_{B_\varrho} \Big|\nabla h^{1+\frac{\tau}{2}}\Big|^2 \eta^2 \, dx\Big)^{\frac{1}{2}} \leqq \max_{B_\varrho} |\nabla \eta| \Big(\int_{B_\varrho} h^{2+\tau} \, dx\Big)^{\frac{1}{2}}. \tag{5.4.13}$$

Put now $B_l = B_{R(1+2^{-l})}(x^0)$, $l = 0, 1, 2, \ldots$, $\tau_l = 2[k^l - 1]$, where $k = \dfrac{n}{n-2}$ for $n \geqq 3$ and $k > 1$ for $n = 2$. Let $\eta_l \in \mathscr{D}(B_l)$, $\eta_l(x) = 1$ in B_{l+1} be chosen in such a way that $|\nabla \eta_l| \leqq 2^{l+1} R^{-1}$. Let us follow the case of $n \geqq 3$. We have from the imbedding theorem and from (5.4.13) that

$$\begin{aligned} \Big(\int_{B_{l+1}} h^{2k^{l+1}} \, dx\Big)^{\frac{1}{k}} &\leqq \Big(\int_{B_l} (h^{k^l}\eta_l)^{2k} \, dx\Big)^{\frac{1}{k}} \leqq c_2 R_l^{\frac{n}{k} - n + 2} \int_{B_l} |\nabla(h^{k^l}\eta_l)|^2 \, dx \\ &\leqq 2c_2 R_l^{\frac{n}{k} - n + 2} \Big[\int_{B_l} |\nabla h^{k^l}|^2 \eta_l^2 \, dx + \int_{B_l} h^{2k^l} |\nabla \eta_l|^2 \, dx\Big] \\ &\leqq c_3 4^l R_l^{\frac{n}{k} - n} \int_{B_l} h^{2k^l} \, dx. \end{aligned} \tag{5.4.14}$$

Put now

$$\Phi_l = \Big(R_l^{-n} \int_{B_l} h^{2k^l} \, dx\Big)^{\frac{l}{2k^l}}; \tag{5.4.15}$$

we get the relation

$$\Phi_{l+1} \leqq c_3^{\frac{1}{2k^l}} 2^{\frac{1}{k^l}} \Phi_l, \tag{5.4.16}$$

thus

$$\Phi_{l+1} \leqq c_5 \Phi_0 \leqq c_5 R^{-\frac{n}{2}} \Big(\int_{B_{2R}} u^2 \, dx\Big)^{\frac{1}{2}}. \tag{5.4.17}$$

First, we let $m \to \infty$ in (5.4.17) and then $l \to \infty$.

5.4.18. LEMMA. *Let $u \in W^{1,2}(B_\varrho(x^0)) \cap L^\infty(B_\varrho(x^0))$ be a solution to the equation* (5.4.2). *Let $\phi \in C^2(\mathbb{R}^1)$, $\phi''(t) \geqq 0$. Then*

(5.4.19)
$$\int_{B_\varrho(x^0)} \left[\phi''(u)\, a_{ij} \frac{\partial u}{\partial x_i} \frac{\partial u}{\partial x_j} \xi + a_{ij} \frac{\partial v}{\partial x_j} \frac{\partial \xi}{\partial x_i}\right] = 0,$$

where $\xi \in W_0^{1,2}(B_\varrho(x^0)) \cap L^\infty(B_\varrho(x^0))$, $v(x) = \phi(u(x))$, $\xi \geqq 0$. Hence v is a weak subsolution.

Proof. Put $\Psi = \phi'(u)\, \xi$. Then

$$0 = \int_{B_\varrho(x^0)} a_{ij}\phi''(u) \frac{\partial u}{\partial x_i} \frac{\partial u}{\partial x_j} \xi \,\mathrm{d}x$$

$$+ \int_{B_\varrho(x^0)} a_{ij}\phi'(u) \frac{\partial \xi}{\partial x_i} \frac{\partial u}{\partial x_j} \mathrm{d}x \geqq \int_{B_\varrho(x^0)} a_{ij} \frac{\partial \xi}{\partial x_i} \frac{\partial v}{\partial x_j} \mathrm{d}x.$$

5.4.20. THEOREM (*of the Harnack type*). *Let $u \in W^{1,2}(B_{4R}(x^0))$, $u \geqq 0$, be a solution to* (5.4.2). *Let $\|a_{ij}\|_{L^\infty(B_{4R}(x^0))} \leqq \mu$, $a_{ij}\xi_i\xi_j \geqq \alpha|\xi|^2$, $\alpha > 0$. Let $S = \{x \in B_{4R}(x^0) \mid u(x) \geqq 1\}$, $|S| \geqq c_1|B_{4R}(x^0)|$. Then*

(5.4.21)
$$u(x) \geqq c_2(\alpha, \mu, c_1) > 0$$

in $B_R(x^0)$.

Proof. First, it follows from 5.4.9 that $u \in L^\infty_{\text{loc}}(B_{4R}(x^0))$. Let $2 < k < 4$ be such that $|B_{4R} - B_{kR}| = \dfrac{c_1}{2}|B_{4R}|$. Then $|S \cap B_{kR}| \geqq \dfrac{c_1}{2}|B_{kR}|$. Let $0 < \varepsilon < 1$ and put $F(u) = [-\log(u + \varepsilon)]^+ = v$. There exists a sequence $F_t \in C^{(2)}(\mathbb{R}_1^+)$, such that $|F_t'| \leqq c_2(\varepsilon)$, $F_t'' \geqq 0$, $F_t(s) = F(s)$ for $|s - 1 + \varepsilon| > \dfrac{1}{t}$. It follows from (5.4.19) for $k < k' < 4$ that

(5.4.22)
$$0 = \int_{B_{k'R}} \left[a_{ij}F_t'' \frac{\partial u}{\partial x_i} \frac{\partial u}{\partial x_j} \xi + a_{ij} \frac{\partial v_t}{\partial x_j} \frac{\partial \xi}{\partial x_i}\right] \mathrm{d}x.$$

Let $\varepsilon_n \to 0$, $\varepsilon_n > 0$, be a sequence such that $|\{x \mid u(x) = 1 - \varepsilon_n\}| = 0$. Let ε_n be fixed and let $t \to \infty$. Put $M_t = \left\{x \in B_{k'R}(x^0)\, |u(x) - 1 + \varepsilon_n| \leqq \dfrac{1}{t}\right\}$. It follows from (5.4.22) that

(5.4.23)
$$0 \geqq \int_{B_{k'R} \setminus M_t} a_{ij} \frac{\partial v}{\partial x_i} \frac{\partial v}{\partial x_i} \xi \,\mathrm{d}x + \int_{B_{k'R}} a_{ij} \frac{\partial v_t}{\partial x_j} \frac{\partial \xi}{\partial x_i} \mathrm{d}x$$

$$\to \int_{B_{k'R}} a_{ij} \frac{\partial v}{\partial x_i} \frac{\partial v}{\partial x_j} \xi \,\mathrm{d}x + \int_{B_{k'R}} a_{ij} \frac{\partial v}{\partial x_j} \frac{\partial \xi}{\partial x_i} \mathrm{d}x.$$

Let $\xi = \eta^2$, $\eta = 1$ in B_{kR}, $\eta \in \mathscr{D}(B_{k'R})$, $|\nabla\eta| \leqq c_3 R^{-1}$. It follows from (5.4.23) with $\xi = \eta^2$ that

$$\int_{B_{k'R}} \eta^2|\nabla v|^2 \,\mathrm{d}x \leqq c_4 R^{-1} \int_{B_{k'R}} |\nabla v|\eta \,\mathrm{d}x \leqq c_5 R^{-1+\frac{n}{2}} \left(\int_{B_{k'R}} \eta^2|\nabla v|^2 \,\mathrm{d}x\right)^{\frac{1}{2}},$$

hence

(5.4.24)
$$\int_{B_{kR}} |\nabla v|^2 \, dx \leqq c_6 R^{n-2}.$$

But $v = 0$ in $S \cap B_{kR}$, thus it follows from the next lemma that

(5.4.25)
$$\mathbb{R}^{-2} \int_{B_{kR}} v^2 \, dx \leqq c_7 \int_{B_{kR}} |\nabla v|^2 \, dx.$$

But v is a subsolution as follows from (5.4.23), hence the combination of (5.4.10), (5.4.24), and (5.4.25) leads to

$$\sup_{x \in B_R} v(x) \leqq c_8 .$$

Letting $\varepsilon_n \to 0$, we get the result.

5.4.26. LEMMA. *Let $M = \{v \in W^{1,2}(B_1(O)) \mid v = 0$ on S with $|S| \geqq c_1 > 0\}$ (S depends on v but not on c_1). Then $\int_{B_1(O)} v^2 \, dx \leqq c_2 \int_{B_1(O)} |\nabla v|^2 \, dx$.*

Proof. Let us suppose that the lemma is not true. Then there exist v_n, $\|v_n\|_{W^{1,2}(B_1)} = 1$, such that $v_n = 0$ on S_n, $|S_n| \geqq c_1$, and such that

(5.4.27)
$$\int_{B_1} v_n^2 \, dx > n \int_{B_1} |\nabla v_n|^2 \, dx.$$

From the compactness of the imbedding $W^{1,2}(B_1) \subsetneq L^2(B_1)$ it follows $v_n \to v_0$ in $L^2(B_1)$ and in virtue of (5.4.27) it is $\int_{B_1} |\nabla v_0|^2 \, dx = 0$, thus $v_0 = |B_1|^{-\frac{1}{2}}$. On the other hand,

(5.4.28)
$$0 \leftarrow \int_{B_1} |v_n - v_0|^2 \, dx \geqq \int_{S_n} |v_0|^2 \, dx \geqq c_1 |B_1|^{-1}$$

which is impossible.

5.4.29. THEOREM (*of the Liouville type*). *Let $u \in W^{1,2}_{\text{loc}}(\mathbb{R}^n)$ be a bounded solution to (5.4.2) with $a_{ij} \in L^\infty(\mathbb{R}^n)$, $a_{ij}\xi_i\xi_j \geqq \alpha|\xi|^2$, $\alpha > 0$. Then $u = \text{const}$.*

Proof. Put $m^*(4R) = \inf_{x \in B_{4R}(O)} u(x)$, $M^*(4R) = \sup_{x \in B_{4R}(O)} u(x)$ and let $m^* \leqq \bar{m} \leqq M^*$ be such that $|\{x \in B_{4R} \mid u(x) > \bar{m}\}| \leqq \dfrac{|B_{4R}|}{2}$, $|\{x \in B_{4R} \mid u(x) < \bar{m}\}| \leqq \dfrac{|B_{4R}|}{2}$. Let us consider the case of $m^* < \bar{m} < M^*$ the rest is left to the reader. Actually, the functions $\dfrac{M^* - u(x)}{M^* - \bar{m}}$ and $\dfrac{u(x) - m^*}{\bar{m} - m^*}$ satisfy all the conditions of Theorem 5.4.20 with $c_1 = \frac{1}{2}$. Hence

$$\frac{M^* - u(x)}{M^* - \bar{m}} \geqq c_2 \Rightarrow u(x) \leqq (1 - c_2)\, M^* + c_2 \bar{m}$$

in $B_R(x^0)$ and, similarly, $u(x) \geqq (1 - c_2)\, m^* + c_2 \bar{m}$. If we write $\omega(\varrho) = M^*(\varrho) - m^*(\varrho)$, we get

(5.4.30)
$$\omega(R) \leqq (1 - c_2)\, \omega(4R) \leqq (1 - c_2)\, \omega(\infty),$$

hence

(5.4.31)
$$\omega(\infty) \leqq (1 - c_2)\, \omega(\infty).$$

Let us remark that it follows just immediately from Theorem 5.4.20 that the solutions to the equation (5.4.2) are Hölder-continuous, see [2] and the following:

5.4.32. THEOREM. *Let $u \in W^{1,2}(\Omega)$, $\Omega \subset \mathbb{R}^n$ be a bounded domain, and u be a solution to* (5.4.2) *with $L^\infty(\Omega)$ coefficients, satisfying* (5.4.1). *Then $u \in C^\mu(\Omega)$; more precisely, we have for $B_{2R}(x^0) \subset \Omega$, $\varrho \leqq R$, that*

$$\left|\sup_{B_\varrho(x^0)} u(x) - \inf_{B_\varrho(x^0)} u(x)\right| \leqq c\varrho^\mu R^{-\mu-\frac{n}{2}}\Big(\int_{B_{2R}(x^0)} u^2(x)\,\mathrm{d}x\Big)^{\frac{1}{2}}. \tag{5.4.33}$$

Proof. Let $F(u) = |u|$. The same argument as in the proof Theorem 5.4.20 proves that $|u|$ is a subsolution. We thus have (5.4.10) and $u \in L^\infty_{\mathrm{loc}}(\Omega)$. Now we shall proceed as in the proof of Theorem 5.4.29. The inequality (5.4.30) together with (5.4.10) give

$$\omega(R4^{-k}) \leqq (1 - c_2)^k\,\omega(R) \leqq c_3(1 - c_2)^k R^{-\frac{n}{2}} \int_{B_{2R}} u^2\,\mathrm{d}x. \tag{5.4.34}$$

Let $\mu > 0$ be such that $4^\mu h = 1$ for $1 - c_2 = h$. Let $R\,4^{-k+1} \leqq \varrho \leqq R\,4^{-k}$. It follows from (5.4.34) that $\omega(\varrho) \leqq \omega(4^{-k}R) \leqq c_4 R^{-\mu}\varrho^\mu R^{-\frac{n}{2}} \int_{B_{2R}(x^0)} u^2\,\mathrm{d}x$. But $\varrho^{-\frac{n}{2}}\Big(\int_{B_\varrho(x^0)} \Big[u(x) - |B_\varrho(x^0)|^{-1} \int_{B_\varrho(x^0)} u\,\mathrm{d}y\Big]^2 \mathrm{d}x\Big)^{\frac{1}{2}} \leqq \omega(\varrho)$ and we get $u \in C^\mu(\Omega)$ in virtue of Theorem 2.6.20.

5.5. Estimate of sup $|\nabla u(x)|$ in virtue of the bounded slope condition

The study of the scalar case for the nonlinear equation

$$-\frac{\partial}{\partial x_i}[a_i(\nabla u)] = 0 \quad \text{in } \Omega, \qquad u = \phi \quad \text{on } \partial\Omega, \tag{5.5.1}$$

can be based on a very elegant estimate of the $W^{1,\infty}(\Omega)$ norm of the solution. We refer to the lecture notes by G. STAMPACCHIA [47] where more details can be found. This method leads also to the existence theorems without any growth conditions, thus also minimal surface problems can be attacted by this method.

Let us mention also the works by G. STAMPACCHIA [54], P. HARTMAN, G. STAMPACCHIA [55], P. HARTMAN [56]. See also [4] and [7].

So let $\Omega \subset \mathbb{R}^n$ be a bounded domain with a Lipschitz boundary and let $u \in W^{1,p}(\Omega)$, $1 < p < \infty$, be a weak solution to (5.5.1), i.e., $\forall v \in W^{1,p}_0(\Omega)$ we have

$$\int_\Omega a_i(\nabla u)\frac{\partial \phi}{\partial x_i}\,\mathrm{d}x = 0. \tag{5.5.2}$$

Let us suppose that $a_i(\eta)$ are continuous and their growth is

$$|a_i(\eta)| \leqq c(1 + |\eta|^{p-1}). \tag{5.5.3}$$

5.5.4. DEFINITION. We say that $u \in W^{1,p}(\Omega)$ satisfies the bounded slope condition (BSC) with a constant K_0 on $\partial\Omega$ if $u \in C(\partial\Omega)$ and if $\forall x^0 \in \partial\Omega$, there exist two linear functions $\pi^{\pm}(x) = \alpha_j^{\pm}(x_j - x_j^0) + u(x^0)$ such that for $x \in \partial\Omega$ it is

(5.5.5) $$\pi^-(x) \leqq u(x) \leqq \pi^+(x), \qquad \alpha_j^+\alpha_j^+ \leqq K_0^2, \qquad \alpha_j^-\alpha_j^- \leqq K_0^2.$$

5.5.6. REMARK. It is true that if u on $\partial\Omega$ is not a restriction of a linear function, then the BSC implies that Ω is convex.

In the next we assume the condition of strict monotony

(5.5.7) $$[a_i(\eta') - a_i(\eta)]\,(\eta_i' - \eta_i) > 0 \quad \text{for } \eta' \neq \eta.$$

5.5.8. THEOREM (*Comparison*). *Let $u, v \in W^{1,p}(\Omega)$ be two solutions of* (5.5.2) *and let $u \leqq v$ on $\partial\Omega$. Let* (5.5.7) *hold. Then $u \leqq v$ almost everywhere in Ω.*

Proof. Put $w = \min(u, v)$. Then $w \in W^{1,p}(\Omega)$, $w - u = 0$ on $\partial\Omega$, hence

(5.5.9)
$$\int_\Omega a_i(\nabla u)\left(\frac{\partial w}{\partial x_i} - \frac{\partial u}{\partial x_i}\right) dx = 0 = \int_{\{x \in \Omega \mid u(x) \geqq v(x)\}} a_i(\nabla u)\left(\frac{\partial v}{\partial x_i} - \frac{\partial u}{\partial x_i}\right) dx.$$

In the same way, if we put $\bar{w} = \max(u, v)$ then

(5.5.10) $$\int_{\{x \in \Omega \mid u(x) \geqq v(x)\}} a_i(\nabla v)\left(\frac{\partial u}{\partial x_i} - \frac{\partial v}{\partial x_i}\right) dx = 0,$$

hence

$$\int_{\{x \in \Omega \mid u(x) \geqq v(x)\}} [a_i(\nabla u) - a_i(\nabla v)]\left(\frac{\partial u}{\partial x_i} - \frac{\partial v}{\partial x_i}\right) dx = 0,$$

so by virtue of (5.5.7) we have $\nabla\omega = 0$ in Ω and $\omega = 0$ on $\partial\Omega$ for $\omega = \max(u, v) - v$. Thus $\omega = 0$.

5.5.11. THEOREM. *Let $u \in W^{1,p}(\Omega)$ be a weak solution to* (5.5.2). *Then*

$$\inf_{x \in \partial\Omega} u(x) \leqq \inf_{x \in \Omega} u(x) \leqq \sup_{x \in \Omega} u(x) \leqq \sup_{x \in \partial\Omega} u(x).$$

For the proof, see 5.4.5.

5.5.12. REMARK. Let $\Omega \subset \mathbb{R}^n$ be a convex domain with a Lipschitz boundary. Let $\Omega_{-\tau} = \{x \in \mathbb{R}^n, x + \tau \in \Omega\}$. Then $\Omega \cap \Omega_{-\tau}$ is a convex domain with a Lipschitz boundary.

To see this, let $x^0 \in \Omega \cap \Omega_{-\tau}$. Let us describe the boundary of $\Omega \cap \Omega_{-\tau}$ by $\operatorname{dist}(x^0, \partial(\Omega \cap \Omega_{-\tau}))$, which is clearly a Lipschitz continuous function defined on $\partial B_1(x^0)$.

5.5.13. THEOREM. *Let Ω be a convex domain with a Lipschitz boundary. Let $u \in W^{1,p}(\Omega)$ be a weak solution to the equation* (5.5.2) *and let the condi-*

tions (5.5.3), (5.5.7) *be satisfied. Let the trace of u satisfy the BSC with a constant* K_0. *Then*

$$|\nabla u(x)| \leqq K_0 \tag{5.5.14}$$

almost everywhere in Ω.

Proof. Let $x^0 \in \partial\Omega$. The functions π^+, π^- corresponding to the point x^0 are also solutions to (5.5.2), hence in virtue of Theorem 5.5.8, the relation (5.5.5) holds in Ω and therefore

$$|u(x) - u(x^0)| \leqq K_0|x - x^0|. \tag{5.5.15}$$

Let further $x, y \in \Omega$ and $\tau = x - y$. Now $u(x)$ and $u(x + \tau)$ are solutions to (5.4.2) on $\Omega \cap \Omega_{-\tau}$ and Theorem 5.5.11 implies

$$\sup_{x \in \Omega \cap \Omega_{-\tau}} |u(x + \tau) - u(x)| \leqq \sup_{x \in \partial(\Omega \cap \Omega_{-\tau})} |u(x + \tau) - u(x)|; \tag{5.5.16}$$

in fact, if $\Phi = \sup_{x \in \partial(\Omega \cap \Omega_{-\tau})} |u(x + \tau) - u(x)|$, then it follows from 5.5.8 that $u(x) - \Phi \leqq u(x + \tau) \leqq u(x) + \Phi$ in $\Omega \cap \Omega_{-\tau}$. Since $\partial(\Omega \cap \Omega_{-\tau}) \subset \partial\Omega \cup \partial\Omega_{-\tau}$, then for $x \in \partial(\Omega \cap \Omega_{-\tau})$ we have according to (5.5.15), that

$$|u(x) - u(x + \tau)| \leqq K_0|\tau|, \tag{5.5.17}$$

hence (5.5.16) implies the result.

Chapter 6

Regularity of weak solutions to second order elliptic systems

6.1. Introduction

Let us add some further remarks to the regularity problem that we spoke about in 1.1.

As the reader knows, for the sake of simplicity we are interested in the second order nonlinear elliptic systems

$$-\frac{\partial}{\partial x_i}[a_i^r(x, u, \nabla u)] + a^r(x, u, \nabla u) = f^r \quad \text{in } \Omega, \tag{6.1.1}$$

$r = 1, 2, \ldots, m$, $u = (u_1, u_2, \ldots, u_m)$. We shall suppose throughout this chapter that $a_i^r(x, \xi, \eta)$ and $a^r(x, \xi, \eta)$ are once continuously differentiable with respect to all the variables in $\bar{\Omega} \times \mathbb{R}^m \times \mathbb{R}^{mn}$, where $\Omega \subset \mathbb{R}^n$ is the considered domain.

Also the condition of very strong ellipticity will be preserved:

$$\frac{\partial a_i^r}{\partial \eta_j^s}(x, \xi, \eta)\, \zeta_i^r \zeta_j^s > 0 \quad \text{for } \zeta \neq 0. \tag{6.1.2}$$

With some exceptions, we suppose that our solutions are from the space $[W^{1,\infty}(\Omega)]^m$; the reader can see some conditions for the weak solutions from $W^{1,p}$, under which such estimates are true, in Chapter 5 and can notice that this situation is far from being satisfactory. On the other hand, this situation, as we just mentioned in 1.1., is, in some sense, artificial, and is connected with a simple but neglected fact that a weak solution, as a model, is a process, which has a solution only in the domain of definition of the response functions a_i^r, a^r, which, in fact, is bounded. So "a priori", it has sense to speak only on solutions with bounded gradient. Theoretically, however, the $W^{1,\infty}$ estimates for systems (6.1.1) are not known with the exception of some special results from Chapter 5. This unpleasant situation is much better if we replace the condition $\nabla u \in [L^\infty(\Omega)]^{mn}$ by $\nabla u \in [\mathscr{L}^{2,n}(\Omega)]^{mn}$, see Chapter 2 for the definition. We get such estimates under some reasonable asymptotic conditions

(6.1.3) $$a_i^r(x, \xi, \eta) = A_{ij}^{rs}(x)\,\eta_j^s + A_i^{rs}(x)\,\xi_s + b_i^r(x, \xi, \eta),$$

with A_{ij}^{rs}, A_i^{rs} smooth enough and with $|b_i^r(x, \xi, \eta)| \leqq c < \infty$, for example.

A great deal of what follows in this chapter can be modified if the L^∞ conditions are replaced by the $\mathscr{L}^{2,n}$ ones, see J. DANĚČEK [58].

Let us define the *interior regularity* of the system (6.1.1): let $f \in [W^{1,p/2}(\Omega)]^m$, $p > n$, and let us consider a whole family of weak solutions to (6.1.1) such that

(6.1.4) $$\|u\|_{[W^{1,\infty}(\Omega)]^m} + \|f\|_{[W^{1,p/2}(\Omega)]^m} \leqq \mu < \infty .$$

So we have for u from this family and for every compact set $K \subset \Omega$ that

(6.1.5) $$\|u\|_{\left[C^{1,1-\frac{n}{p}}(K)\right]^m} \leqq c(\mu) < \infty .$$

6.1.7. DEFINITION. *The condition $L(\mathbb{R}^n)$ of the Liouville type is satisfied for the system* (6.1.1) *if $\forall x^0 \in \Omega$, $\forall \xi \in \mathbb{R}^m$, the only weak solutions in $\mathbb{R}^n$ to the system*

(6.1.8) $$-\frac{\partial}{\partial x_i}\,[a_i^r(x^0, \xi, \nabla v)] = 0$$

with bounded gradient $|\nabla v| \leqq c < \infty$ are polynomials of at most first degree.

One of the main results in this chapter is the assertion that the regularity in the sense (6.1.4), (6.1.5) follows from $L(\mathbb{R}^n)$ and that if the special systems

(6.1.9) $$-\frac{\partial}{\partial x_i}\,[a_i^r(\nabla u)] + a^r(x, u, \nabla u) = f^r,$$

are considered then $L(\mathbb{R}^n)$ is also necessary.

Let us mention the well-known fact, that if $u \in [C^{1,\mu}(\Omega)]^m$ is a weak solution to the system (6.1.1) then the solution is from $C^{l,\alpha}$, $l \geqq 2$, provided that the coefficients are l-times continuously differentiable and f^r is from $C^{l-2,\alpha}$. This follows from the theory of linear systems, see for example S. AGMON, A. DOUGLIS, and L. NIRENBERG [59]. Moreover, if the coefficients are analytic, the solution is analytic (provided that the solution is from $C^{1,\alpha}$), see [2].

We mentioned in 1.1. that the condition $L(\mathbb{R}^n)$ is not always satisfied. From this point, one of the most important counter-examples is due to E. DE GIORGI [60] who presented a linear system (6.1.1) with L^∞ coefficients and with an unbounded weak solution. A nonlinear modification of this example was constructed by E. GIUSTI and M. MIRANDA [61]. It can be written in the form

(6.1.10) $$-\frac{\partial}{\partial x_i}\left[A_{ij}^{rs}(u)\,\frac{\partial u_s}{\partial x_j}\right] = 0, \quad n \geqq 3,$$

where $u \in W_{\mathrm{loc}}^{1,2}(\mathbb{R}^n)$, $u \in L^\infty(\mathbb{R}^n)$, $u = \dfrac{x}{|x|}$, $A_{ij}^{rs} = \delta_{rs}\delta_{ij} + \left[\delta_{ri} + \dfrac{4}{n-2}\,\dfrac{u_r u_i}{1+|u|^2}\right] \times \left[\delta_{sj} + \dfrac{4}{n-2}\,\dfrac{u_s u_j}{1+|u|^2}\right].$

The system (6.1.10) is a typical quasilinear system anc an analogous theory for a system of this kind can be developed via the Liouville type condition (where bounded solutions in $\mathbb{R}^n$ are considered) as was done by B. KAWOHL [62].

In general, a partial regularity can be proved; see 1 1. and what follows. Here the fundamental work is due to E. GIUSTI [63]. It is also the basis of our considerations which follow. See also the first work on this topic by CH. B. MORREY [63]'. For a counter-example to the systems $-\frac{\partial}{\partial x_i}[a_i^r(\nabla u)] = 0$ in the whole $\mathbb{R}^n$ with the solutions with bounded gradients ($n \geqq 3$) which are not polynomials, see [16], [21] and J. NEČAS, O. JOHN, J. STARÁ [64]. The solutions have the form $\frac{x_i x_j}{|x|} - \frac{1}{n}\delta_{ij}|x|$. The only chance for regularity for $n = 3$ is the potential case, because up to this time, there is no such counter-example for the second order systems (6.1.1) that are Euler's systems. There is an unpublished result of J. NEČAS with one equation cf the 14th order with the solution $|x|^7$ and with Euler's system for $n = 3$. For n large enough, such a system was constructed by J. NEČAS [66].

We reprove once more in these lecture notes, that $L(\mathbb{R}^n)$ is true if $n = 2$ or $m = 1$. So it is a very important question, the question on reasonable conditions that guarantee $L(\mathbb{R}^n)$. Up to this time, it is not much known about it. One way is the condition number approach: we consider the system (6.1.1) and suppose for the sake of simplicity that $\frac{\partial a_i^r}{\partial \eta_j^s} = \frac{\partial a_j^s}{\partial \eta_i^r}$. For all x, ξ, η let

(6.1.11) $$\lambda_1|\zeta|^2 \leqq \frac{\partial a_i^r}{\partial \eta_j^s}(x, \xi, \eta)\, \zeta_i^r \zeta_j^s \leqq \lambda_2|\zeta|^2.$$

If

(6.1.12) $$\frac{\lambda_1}{\lambda_2} > \frac{\sqrt{1 + \frac{(n-2)^2}{n-1}} - 1}{\sqrt{1 + \frac{(n-2)^2}{n-1}} + 1},$$

then $L(\mathbb{R}^n)$ is satisfied. See [21] and a completely different approach in J. NEČAS [67] and S. I. KOŠELEV [68]. A very nice method in this direction is due to O. A. OLEĬNIK, see J. NEČAS, O. A. OLEĬNIK [69] and also J. NEČAS [70], see later.

A completely different approach is due to J. NEČAS, J. STARÁ, R. ŠVARC [71] for the dimension $n = 3$, that concerns also the regularity up to the boundary. They consider Euler's system

(6.1.13) $$-\frac{\partial}{\partial x_i}\left[\frac{\partial F}{\partial \eta_i^r}(\nabla u)\right] = f^r,$$

with

(6.1.14) $$\frac{\partial^2 F}{\partial \eta_i^r \partial \eta_j^s}(\eta)\, \zeta_i^r \zeta_j^s \geqq \nu|\zeta|^2,$$

where $|\eta| \leqq a$ (in the Euclidean norm).

For $|\eta| \leqq a$ let

(6.1.15)
$$\frac{\partial^4 F}{\partial\eta_i^r \partial\eta_j^s \partial\eta_k^t \partial\eta_l^v}(\eta)\,\zeta_i^r\zeta_j^s\zeta_k^t\zeta_l^v \leqq T|\zeta|^4.$$

Thus if $\nu - 3a^2T > 0$, then $u \in [W^{3,2}(\Omega)]^m$, provided that $f^r \in W^{1,2}(\Omega)$ and $u = u^0$ on $\partial\Omega$ (regular enough) with $u^0 \in [W^{3,2}(\Omega)]^m$.

This is applied in [70] to finite elasticity and to the classical Saint-Venant-Kirchhoff law-between strain and stress. Here, we shall apply this result to the same material, but it is shown to be equivalent, in some sense, to the Ogden material[1]), and to the nonlinear Mises-Hencky elasticity as is done in [70].

The regularity of the weak solution to the system (6.1.1) up to the boundary can be formulated similarly via the Liouville type condition:

Let us consider, for the sake of simplicity, the Dirichlet problem in the cylinder $C = \{x \in \mathbb{R}^n,\ x = (x', x_n) \mid |x'| < 1,\ 0 < x_n < 1\}$. The regularity, up to the boundary (up to a part of the boundary) is defined in the following way: let $C_t = \{x \in C \mid |x'| < t,\ 0 < x_n < t\}$, $\Gamma = \{x \in \partial C \mid x_n = 0\}$.

Consider a family of solutions to (6.1.1) such that $u = u^0$ on Γ and that

(6.1.16)
$$\|u\|_{[W^{1,\infty}(C)]^m} + \|f\|_{[W^{1,p/2}(C)]^m} + \|u^0\|_{[W^{2,p}(C)]^m} \leqq \mu < \infty.$$

For $t < 1$ then

(6.1.17)
$$\|u\|_{C^{1,1-n/p}(\bar{C}_t)} \leqq c(t, \mu) < \infty.$$

6.1.18. DEFINITION. *The condition $L(\mathbb{R}^n_+)$ of the Liouville type is satisfied for the system* (6.1.1) *if for $\forall x^0 \in \Gamma$, $\forall \xi \in \mathbb{R}^m$, the only weak solutions v in $\mathbb{R}^n_+$ with the bounded gradient $|\nabla v| \leqq c < \infty$ in $\mathbb{R}^n_+$ to the system* (6.1.1) *such that $v = p$ for $x_n = 0$ and p is a polynomial of at most first degree, are polynomials of at most first degree.*

It is proved in M. Giaquinta, J. Nečas, O. John, J. Stará [72] and we reproduce it in these lecture notes, that $L(\mathbb{R}^n)$ and $L(\mathbb{R}^n_+) \Rightarrow$ the regularity up to the boundary in the sense (6.1.16) and (6.1.17). The necessity is as above for the interior regularity.

Let us close with important open problems:

6.1.19. PROBLEM. Are $L(\mathbb{R}^n)$ and $L(\mathbb{R}^n_+)$ generic?

6.1.20. PROBLEM. Are $L(\mathbb{R}^n_+)$, $L(\mathbb{R}^n_+)$ satisfied, provided that $a_i^r(x^0, \xi, \eta)$ is in η invariant with respect to the rotations of coordinate systems?

6.2. Partial regularity for quasilinear systems and fundamental lemmas

Let b_{ij}^{rs}, $r, s = 1, 2, \ldots, M$, $i, j = 1, 2, \ldots, n$, be constants, $N = b_{ij}^{rs} b_{ij}^{rs}$, and let us assume the very strong ellipticity

(6.2.1)
$$b_{ij}^{rs}\eta_i^r\eta_j^s \geqq \nu|\eta|^2, \quad \nu > 0.$$

[1]) P. Ciarlet pointed me out such a possibility in a private communication.

We have a generalization of a lemma due to S. CAMPANATO [73]:

6.2.2. LEMMA. *Let* $u \in [W^{1,2}_{\mathrm{loc}}(B_1(O))]^M \cap [L^2(B_1(O))]^M$ *be a weak solution to the system*

$$(6.2.3)\qquad -\frac{\partial}{\partial x_i}\left[b_{ij}^{rs}\frac{\partial u_s}{\partial x_j}\right] = 0 \quad \text{in } B_1(O).$$

Then there exists $c = c(N, \nu)$ *such that*

$$(6.2.4)\qquad U(0, \varrho) \leqq c\varrho^2 U(0, 1),$$

where $U(x^0, \varrho) \overset{\text{def}}{=} \varrho^{-n} \int\limits_{B_\varrho(x^0)} |u(x) - (u)^{x^0,\varrho}|^2 \, dx$, $(u)^{x^0,\varrho} \overset{\text{def}}{=} \dfrac{1}{|B_\varrho(x^0)|} \int\limits_{B_\varrho(x^0)} u(x) \, dx$.

Proof. First, we have Cacciopoli's inequality

$$(6.2.5)\qquad \int\limits_{B_{\frac{1}{2}}(O)} |\nabla u|^2 \, dx \leqq c_1 \int\limits_{B_1(O)} |u - (u)^{0,1}|^2 \, dx.$$

In fact, let $\eta(r)$ be an infinitely differentiable function in $\overline{\mathbb{R}}^1_+$ such that $\eta(r) = 1$ for $0 \leqq r \leqq \frac{1}{2}$ and $\eta(r) = 0$ for $r \geqq \frac{3}{4}$, and put $v(x) = (u(x) - (u)^{0,1})\,\eta^2(|x|)$. We have

$$0 = \int\limits_{B_1} b_{ij}^{rs} \frac{\partial v_r}{\partial x_i}\frac{\partial u_s}{\partial x_j}\, dx = \int\limits_{B_1} b_{ij}^{rs} \frac{\partial u_r}{\partial x_i}\frac{\partial u_s}{\partial x_j}\eta^2 \, dx$$
$$+ 2\int\limits_{B_1} b_{ij}^{rs}\eta \frac{\partial \eta}{\partial x_i}(u_r - (u_r)^{0,1})\frac{\partial u_s}{\partial x_j}\, dx,$$

hence (6.2.5) follows. For the solution to (6.2.3), we have as high regularity as we wish, see for example [25]; especially we have

$$(6.2.6)\qquad \max_{x \in B_{\frac{1}{4}}(O)} |\nabla u(x)|^2 \leqq c_2 \int\limits_{B_{\frac{1}{2}}(O)} |\nabla u|^2 \, dx.^{1)}$$

Hence for $t \leqq \frac{1}{4}$ we have $t^{-n} \int\limits_{B_t(O)} |\nabla u|^2 \, dx \leqq c_3 \int\limits_{B_{\frac{1}{2}}(O)} |\nabla u|^2 \, dx$. Finally, we use Poincaré's inequality (see (2.5.2))

$$(6.2.7)\qquad t^{-2}U(0, t) \leqq c_4 \int\limits_{B_t(O)} |\nabla u|^2 \, dx$$

and the result follows.

Let us actually consider the quasilinear system

$$(6.2.8)\qquad -\frac{\partial}{\partial x_i}\left[A_{ij}^{rs}(x, \omega)\frac{\partial \omega_s}{\partial x_j}\right] + A_j^{rs}(x, \omega)\frac{\partial \omega_s}{\partial x_j} = -\frac{\partial g_i^r}{\partial x_i} + g^r,$$
$$r, s = 1, 2, \ldots, M, \quad i, j = 1, 2, \ldots, n,$$

---

[1]) The reader can prove (6.2.6) as an exercise, using Theorem 5.2.1 for all $\dfrac{\partial u}{\partial x_l}$, $l = 1, 2, \ldots, n$, then for $\dfrac{\partial}{\partial x_k}\left(\dfrac{\partial u}{\partial x_l}\right)$, etc. We thus come to the inequality

$$\|\nabla u\|^2_{[W^{l,2}(B_{\frac{1}{4}}(O))]^{mn}} \leqq c\,\|\nabla u\|^2_{[L^2(B_{\frac{1}{2}}(O))]^{mn}}$$

with $2l > n$; using now the imbedding theorem from Chapter 2, the result follows.

in a bounded domain $\Omega \subset \mathbb{R}^n$ and let us assume the coefficients $A_{ij}^{rs}(x, \xi)$, $A_j^{rs}(x, \xi)$ continuous in $\bar{\Omega} \times \mathbb{R}^M$, $g_i^r \in L^p(\Omega)$, $g^r \in L^{p/2}(\Omega)$, $p > n$, and the very strong ellipticity

$$(6.2.9) \qquad A_{ij}^{rs}(x, \xi)\, \zeta_i^r \zeta_j^s > 0 \quad \text{for } \zeta \neq 0, \quad \zeta \in \mathbb{R}^{Mn}.$$

We shall consider weak solutions in $[W^{1,2}(\Omega)]^M \cap [L^\infty(\Omega)]^M$ or, also, only in $[W^{1,2}(\Omega)]^M$ which we shall especially mention.

Let $x^0 \in \Omega$, $R < \operatorname{dist}(x^0, \partial\Omega)$ and so small that the linear system

$$(6.2.10) \qquad -\frac{\partial}{\partial x_i}\left[A_{ij}^{rs}(x, \omega)\frac{\partial w_s}{\partial x_j}\right] + A_j^{rs}(x, \omega)\frac{\partial w_s}{\partial x_j} = -\frac{\partial g_i^r}{\partial x_i} + g^r$$

has a unique solution in $[W_0^{1,2}(B_R(x^0))]^M$; we shall see later that this is true.

Let us decompose the solution ω as

$$(6.2.11) \qquad \omega = v^{x^0,R} + w^{x^0,R};$$

if there will not be the danger of misunderstanding, we shall omit the superscripts x^0, R.

We have the fundamental lemma due to E. Giusti [63]:

6.2.12. LEMMA. *Let ω be a solution to* (6.2.10) *from the family*

$$\|\omega\|_{[L^\infty(\Omega)]^M} + \sum_{r,i} \|g_i^r\|_{L^p(\Omega)} + \sum_r \|g^r\|_{L^{p/2}(\Omega)} \leqq \mu.$$

Let $K \subset \Omega$ be compact. For $\forall\tau$, $0 < \tau < 1$, then there exist $\varepsilon_0 = \varepsilon_0(\mu, \tau, K)$ and $R_0 = R_0(\mu, \tau, K)$ such that for $R \leqq R_0$ and $x^0 \in K$ we have

$$V(x^0, R) \overset{\text{def}}{=} R^{-n} \int_{B_R(x^0)} |v - (v)^{x^0,R}|^2 \,dx < \varepsilon_0^2 \Rightarrow$$

$$(6.2.13) \qquad V(x^0, \tau R) \leqq 2c\tau^2 V(x^0, R),$$

where c is the constant from Lemma 6.2.2, *common for all the coefficients $A_{ij}^{rs}(y, \xi)$, $y \in K$, $\xi \in \mathbb{R}^m$, $|\xi| \leqq \mu$.*

Proof. Let us suppose that the lemma is not true for some τ. Then there exist $x^\nu \in K$, $x^\nu \to x^0$, $\varepsilon_\nu \to 0$, $R_\nu \to 0$, and ω^ν from our family, $\omega^\nu = v^\nu + w^\nu$ and such that $V^\nu(x^\nu, R_\nu) = \varepsilon_\nu^2$ and

$$(6.2.14) \qquad V^\nu(x^\nu, \tau R_\nu) > 2c\tau^2\varepsilon_\nu^2.$$

Put $x = x^\nu + R_\nu y$, $s^\nu(y) = \varepsilon_\nu^{-1}\left[v^\nu(x^\nu + R_\nu y) - (v^\nu)^{x_\nu,R_\nu}\right]$. We have $S^\nu(0, 1) = 1$ and

$$(6.2.15) \qquad S^\nu(0, \tau) = \tau^{-n} \int_{B_\tau(O)} |s^\nu(y) - (s^\nu)^{0,\tau}|^2 \,dy$$

$$= \tau^{-n} R_\nu^{-n} \varepsilon_\nu^{-2} \int_{B_{R_\nu\tau}(x^\nu)} |v^\nu - (v^\nu)^{X_\nu, R_\nu} - \varepsilon_\nu (s^\nu)^{0,\tau}|^2 \,dx \geqq \varepsilon_\nu^{-2} V^\nu(x^\nu, \tau R^\nu) > 2c\tau^2.$$

We can suppose that $s^\nu \rightharpoonup s$ in $[L^2(B_1(O))]^M$ and $\varepsilon_\nu s^\nu(y) \to 0$ almost everywhere in $B_1(O)$. Put $t^\nu(y) \overset{\text{def}}{=} w^\nu(x^\nu + R_\nu y)$. We get from (6.2.10) that

$$(6.2.16) \qquad \int_{B_1(O)} |t^\nu(y)|^2 \,dy \leqq c_1 R_\nu^{2\varkappa}, \quad \varkappa = 1 - \frac{n}{p}.$$

But

(6.2.16)′ $v^\nu(x^\nu + R_\nu y) = (v^\nu)^{x^\nu, R_\nu} + \varepsilon_\nu s^\nu(y) = \omega^\nu(x^\nu + R_\nu y) - t^\nu(y)$.

Since $\|\omega^\nu\|_{[L^\infty(\Omega)]^M} \leqq \mu$ and $t^\nu(y) \to 0$ almost everywhere (which we can suppose) in $B_1(O)$, then we can suppose that $(v^\nu)^{x^\nu, R_\nu} \to \xi$ and thus clearly $|\xi| \leqq \mu$. Therefore $A_{ij}^{rs}(x^\nu + R_\nu y, \omega^\nu(x^\nu + R_\nu y)) \to A_{ij}^{rs}(x^0, \xi)$ almost everywhere in $B_1(O)$ and $A_j^{rs}(x^\nu + R_\nu y, \omega^\nu(x^\nu + R_\nu y)) \to A_j^{rs}(x^0, \xi)$ almost everywhere in $B_1(O)$. But with the coefficients $A_{ij}^{rs}(x^\nu + R_\nu y, \omega^\nu(x^\nu + R_\nu y))$, $A_j^{rs}(x^\nu + R_\nu y, \omega^\nu(x^\nu + R_\nu y))$ we have

$$(6.2.17) \qquad \int_{B_1(O)} \left[A_{ij}^{rs} \frac{\partial s_s^\nu}{\partial y_j} \frac{\partial \varphi_r}{\partial y_i} + R_\nu A_j^{rs} \frac{\partial s_s^\nu}{\partial y_j} \varphi_r\right] \mathrm{d}y = 0$$

and the estimate (6.2.5) for $B_t(O)$, $0 < t < 1$, and $B_1(O)$ gives that $s^\nu \in [W^{1,2}(B_t(O))]^M$ and that for every $t < 1$, the norms are bounded here. We can thus suppose that $s^\nu \rightharpoonup s$ in $[W^{1,2}(B_t(O))]^M$ for every $t < 1$. This implies that s is a solution to

$$(6.2.18) \qquad \int_{B_1(O)} A_{ij}^{rs}(x^0, \xi) \frac{\partial s_s}{\partial y_j} \frac{\partial \varphi_r}{\partial y_i} \mathrm{d}y = 0.$$

In virtue of (6.2.15), taking into account that $s^\nu \to s$ in $[L^2(B_t(O))]^M$, we get

$$(6.2.19) \qquad S(0, \tau) \geqq 2c\tau^2,$$

thus $S \neq 0$ and because $S(0, 1) \leqq 1$, we finally obtain

$$(6.2.20) \qquad S(0, \tau) \geqq 2c\tau^2 S(0, 1),$$

which contradicts (6.2.4).

Let us immediately give an easy version of Lemma 6.2.12 which will serve us in partial regularity results without supposing $\omega \in [L^\infty(\Omega)]^M$. For more details on this subject, see [63], [6] and the work by M. GIAQUINTA [74].

So let us consider the systems (6.2.8) with the coefficients satisfying the inequalities

$$(6.2.21) \qquad \sum_{r,i,s,j} |A_{ij}^{rs}| + \sum_{r,s,j} |A_j^{rs}| \leqq c < \infty,$$

$$(6.2.22) \qquad A_{ij}^{rs} \zeta_i^r \zeta_j^s \geqq \nu |\zeta|^2, \quad \nu > 0,$$

and with $\omega \in [W^{1,2}(\Omega)]^M$ only. Let us also suppose that the coefficients $A_{ij}^{rs}(x, \xi)$ are uniformly continuous in $\bar{\Omega} \times \mathbb{R}^M$.

We have the following

6.2.23. LEMMA. *Let ω be the solutions to (6.2.8) from $[W^{1,2}(\Omega)]^M$, provided that the coefficients A_{ij}^{rs}, A_j^{rs} satisfy (6.2.21), (6.2.22) and A_{ij}^{rs} are uniformly continuous in $\bar{\Omega} \times \mathbb{R}^M$. Let these solutions be from the family*

$$(6.2.24) \qquad \sum_{r,i} \|g_i^r\|_{L^p(\Omega)} + \sum_r \|g^r\|_{L^{p/2}(\Omega)} \leqq \mu.$$

Let $K \subset \Omega$ be compact. Then the assertion of Lemma 6.2.12 is valid with a constant common for all the coefficients $A_{ij}^{rs}(y, \xi)$, $y \in K$, $\xi \in \mathbb{R}^m$.

Proof. We do word by word the same proof as for Lemma 6.2.12.

6.2.25. LEMMA. *Let the conditions of Lemma* 6.2.12 *or of Lemma* 6.2.23 *be satisfied. Let*

(6.2.26) $$\lim_{R\to 0} R^{-n} \int_{B_R(x^0)} |v - (v)^{x^0, R}|^2 \, dx = 0$$

uniformly with respect to the families of solutions in question and with respect to x^0 *from K. Then*

(6.2.27) $$\|\omega\|_{[C^{0,1-n/p}(K)]^M} \leqq c < \infty .$$

Proof. Let c be the constant from (6.2.13) and c_1 that from (6.2.16). Choose τ such that $\sqrt{2c}\sqrt{\tau} \leqq \frac{1}{2}$, $\sqrt{2c_1}\, \tau^{n/p} < 1$, $0 < \tau < 1$. It follows from Lemma 6.2.12 or 6.2.23 that there exist ε_0, R_1 with the above mentioned properties.

We have

(6.2.28) $$V^{\tau R_1}(x^0, \tau R_1) \leqq \tau^{-n} R_1^{-n} \int_{B_{\tau R_1}(x^0)} |v^{x^0, \tau R_1}(x) - (v^{x^0, R_1})^{x^0, \tau R_1}|^2 \, dx$$

$$= \tau^{-n} R_1^{-n} \int_{B_{\tau R_1}(x^0)} |v^{x^0, R_1}(x) + w^{x^0, R_1}(x) - w^{x^0, R_1}(x) - (v^{x^0, R_1})^{x^0, \tau R_1}|^2 \, dx .$$

Therefore (6.2.13) and (6.2.16) imply

(6.2.29) $$\sqrt{V^{\tau R_1}(x^0, \tau R_1)} \leqq \sqrt{2c}\, \tau \sqrt{V^{R_1}(x^0, R_1)} + c_1^{\frac{1}{2}} \tau^{-\frac{n}{2}} R_1^{1-\frac{n}{p}} \left[1 + \tau^{1-\frac{n}{p}+\frac{n}{2}}\right].$$

Choose R_1 so small that

$$c_1^{\frac{1}{2}} \tau^{-\frac{n}{2}} R_1^{1-\frac{n}{p}} \left[1 + \tau^{1-\frac{n}{p}+\frac{n}{2}}\right] < \frac{\varepsilon_0}{2} .$$

Then it follows from (6.2.29) that $V^{\tau R_1}(x^0, \tau R_1) < \varepsilon_0^2$ and we can reiterate the process: for an integer $k \geqq 1$ we have

(6.2.30) $$V^{\tau^{k-1} R_1}(x^0, \tau^k R_1) \leqq 2c\tau^2 V^{\tau^{k-1} R_1}(x^0, \tau^{k-1} R_1).$$

Hence we get

(6.2.31) $$\sqrt{\Omega(x^0, \tau^k R_1)} \leqq (\tau^k R_1)^{-\frac{n}{2}} \left(\int_{B_{\tau^k R_1}(x^0)} |\omega - (v^{x^0, \tau^{k-1} R_1})^{x^0, \tau^k R_1}|^2 \, dx \right)^{\frac{1}{2}}$$

$$\leqq \sqrt{2c}\, \tau \sqrt{V^{\tau^{k-1} R_1}(x^0, \tau^{k-1} R_1)} + \gamma \tau^{k\left(1 - \frac{n}{p}\right)}.$$

We thus obtain

(6.2.32) $$\sqrt{\Omega(x^0, \tau^k R_1)} \leqq \gamma_1 \tau^{k\left(1 - \frac{n}{p}\right)}.$$

Let now $\varrho \leqq R_1$ and k be an integer such that $\tau^{k+1} R_1 < \varrho \leqq \tau^k R_1$. We get

(6.2.33) $$\tau^n \Omega(x^0, \varrho) \leqq \left(\frac{\varrho}{R_1 \tau^k}\right)^n \Omega(x^0, \varrho) \leqq \Omega(x^0, \tau^k R_1) \leqq \gamma_1^2 \tau^{2k\left(1 - \frac{n}{p}\right)}$$

$$\leqq \gamma_1^2 \left(\frac{\varrho}{R_1 \tau}\right)^{2\left(1 - \frac{n}{p}\right)}.$$

We can suppose that $K = \bar{\Omega}_1$ where Ω_1 is a domain with a Lipschitz boundary. We use Theorem 2.6.20.

We get also the following modification of Lemma 6.2.25:

6.2.34. LEMMA. *Let ω be a solution to the system* (6.2.8) *and* $\omega \in [W^{1,2}(\Omega)]^M \cap [L^\infty(\Omega)]^M$ *or* $\omega \in [W^{1,2}(\Omega)]^M$. *In the latter case, assume* (6.2.21), (6.2.22), *and the uniform continuity of* $A_{ij}^{rs}(x, \xi)$ *in* $\bar{\Omega} \times \mathbb{R}^M$. *If at some point* $x^0 \in \Omega$

$$\lim_{R\to 0} R^{-n} \int_{B_R(x^0)} |\omega(x) - (\omega)^{x^0, R}|^2 \, dx = 0, \tag{6.2.35}$$

then

$$\omega \in C^{0,1-\frac{n}{p}}(B_{R_0}(x^0)) \quad \text{for some } R_0 > 0. \tag{6.2.36}$$

Proof. In virtue of (6.2.16), it follows with τ, ε^0, and R_1 from Lemma 6.2.25, that for some $B_{R_0}(x^0)$ we have

$$R^{-n} \int_{B_R(y)} |v - (v)^{x^0, R}|^2 \, dx < \varepsilon_0^2 \quad \text{if } R \leqq R_2 \leqq R_1, \quad y \in B_{R_0}(x^0).$$

In fact, it follows immediately from Lemma 6.2.34 that there exists a relatively closed set $M \subset \Omega$ with $|M| = 0$ such that every solution ω in question is from $C^{0,1-n/p}(\Omega \setminus M)$, because (6.2.35) is valid almost everywhere in Ω. But it follows immediately from Poincaré's inequality that

$$R^{-n} \int_{B_R(x^0)} |\omega(x) - (\omega)^{x^0, R}|^2 \, dx \leqq cR^{2-n} \int_{B_R(x^0)} |\nabla\omega|^2 \, dx, \tag{6.2.37}$$

which gives the evidence that the set M is "smaller". Before this, let us recall

6.2.38. DEFINITION. *Let* $E \subset \mathbb{R}^n$ *and let* $E \subset \bigcup_{i=1}^{\infty} F_i$, *where* F_i *are open sets of* diam $F_i < \delta$. *For* $0 \leqq \sigma$, *define* $H_\sigma^{(\delta)}(E) = \inf \left\{ |B_1(O)| \, 2^{-\sigma} \sum_{i=1}^{\infty} (\operatorname{diam} F_i)^\sigma \right\}$ *over such coverings. Put* $H_\sigma(E) = \lim_{\delta\to 0} H_\sigma^{(\delta)}(E)$. H_σ *is the σ-dimensional Hausdorf measure.*

6.2.39. DEFINITION. *Let* $M \subset \mathbb{R}^n$ *and let* F *be a family of* $B_\varrho(x)$, $x \in M$, *such that* $\forall \delta > 0$, $y \in M$, *there exists* $y \in B_\varrho(x) \in F$ *and* $|B_\varrho(x)| < \delta$. *Such a covering is called the Vitali covering.*

The following statement was proved by S. BANACH (for the proof, see also [6] or [74]):

6.2.40. PROPOSITION. *Let* $M \subset \mathbb{R}^n$ *and let* F *be a Vitali covering of* M. *Then there exist* $B_{\varrho_i}(x^i) \in F$, $i = 1, 2, \ldots$, *such that* $B_{\varrho_i}(x^i) \cap B_{\varrho_j}(x^j) = \emptyset$ *for* $i \neq j$ *and* $M \subset \bigcup_{i=1}^{\infty} B_{5\varrho_i}(x^i)$.

6.2.41. THEOREM. *Let* $\omega \in [W^{1,2}(\Omega)]^M \cap [L^\infty(\Omega)]^M$ *or* $\omega \in [W^{1,2}(\Omega)]^M$ *be a solution to the system* (6.2.8) *such that* $\sum_{r,i} \|g_i^r\|_{L^p(\Omega)} + \sum_r \|g^r\|_{L^{p/2}(\Omega)} < \infty$, $p > n$. *We suppose the continuity of* A_{ij}^{rs}, A_j^{rs} *in* $\Omega \times \mathbb{R}^M$ *in the former case or also the uniform continuity of* A_{ij}^{rs} *for* $\bar{\Omega} \times \mathbb{R}^M$ *and* (6.2.9) *or* (6.2.21), (6.2.22), *respectively. Then there exists a relatively closed set* M *in* Ω *such that* $H_{n-2}(M) = 0$ *and that* $\omega \in C^{0,1-n/p}(\Omega \setminus M)$.

Proof. Let K be a compact set in Ω, ε_0, R_0 be from Lemma 6.2.12 or 6.2.23 for some $0 < \tau < 1$. Let $M \overset{\text{def}}{=} \left\{x \in K \mid \overline{\lim_{R\to 0}} R^{-n+2} \int_{B_R(x)} |\nabla\omega|^2 \, dy \geqq 2\varepsilon_0^2\right\}$. Let F be the family of all the balls $B_R(x)$ with $x \in M$, $R \leqq R_1 \leqq R_0$, and such that $R^{-n+2} \int_{B_R(x)} |\nabla\omega|^2 \, dx \geqq \varepsilon_0^2$. Since this is a Vitali covering of M, there exist $B_{R_i}(x_i)$ from Proposition 6.2.40. We have

$$R_1^2 \int_{\bigcup_{i=1}^{\infty} B_{R_i}(x^i)} |\nabla\omega|^2 \, dx \geqq \varepsilon_0^2 \sum_{i=1}^{\infty} R_i^n, \tag{6.2.42}$$

so

$$\left|\bigcup_{i=1}^{\infty} B_{R_i}(x^i)\right| \leqq \frac{R_1^2}{\varepsilon_0^2} \int_{\Omega} |\nabla\omega|^2 \, dx. \tag{6.2.43}$$

On the other hand,

$$\sum_{i=1}^{\infty} R_i^{n-2} \leqq \frac{1}{\varepsilon_0^2} \int_{\bigcup_{i=1}^{\infty} B_{R_i}(x^i)} |\nabla\omega|^2 \, dx. \tag{6.2.44}$$

This can be done for R_1 arbitrarily small, thus $H_{n-2}(M) = 0$.

6.2.45. REMARK. Using the result mentioned first in 4.4.29, the assumptions of Theorem 6.2.41 imply $\omega \in [W^{1,p}_{\text{loc}}(\Omega)]^M$. So we get, as above, that $H_{n-p}(M) = 0$, where $p > 2$.

6.2.46. PROBLEM. Let $\text{Dim}_\alpha(M) \overset{\text{def}}{=} \inf\{\alpha \mid H_\alpha(M) = 0\}$. Under the assumptions of Theorem 6.2.41, is $\text{Dim}_{n-3}(M) = 0$ for $n \geqq 3$?

6.3. Partial regularity for nonlinear elliptic systems and regularity in the interior

Let us consider, as usual, the system

$$-\frac{\partial}{\partial x_i}[a_i^r(x, u, \nabla u)] + a^r(x, u, \nabla u) = -\frac{\partial f_i^r}{\partial x_i} + f^r, \quad r = 1, 2, \ldots, m; \tag{6.3.1}$$

we suppose that $u \in [W^{1,\infty}(\Omega)]^m$, $f_i^r \in W^{1,p}(\Omega)$, $f^r \in W^{1,\frac{p}{2}}(\Omega)$. We shall consider also only weak solutions in $[W^{1,2}(\Omega)]^m$ but we come later to this. We suppose that the coefficients a_i^r, a^r are once continuously differentiable in $\bar{\Omega} \times \mathbb{R}^n \times \mathbb{R}^{mn}$ and assume the condition of very strong ellipticity

$$\frac{\partial a_i^r}{\partial \eta_j^s}(x, \xi, \eta)\, \zeta_i^r \zeta_j^s > 0 \quad \text{for } \zeta \neq 0. \tag{6.3.2}$$

We prove first that the condition $L(R^n)$ defined in 6.1.7 is necessary in the sense explained below. We follow the ideas of J. NEČAS [65] and of [62]; in fact this trick can be found also, for example, in the work by S. I. KRUŽKOV [74].

6.3.3. THEOREM. *Let $\Omega \subset \mathbb{R}^n$ be a bounded domain and $\forall x^0 \in \Omega$, $\forall \xi \in \mathbb{R}^m$, let the family of the solutions u to*

$$-\frac{\partial}{\partial x_i}[a_i^r(x^0, \xi, \nabla u)] = 0 \tag{6.3.4}$$

in $B_1(O)$, such that

$$\|u\|_{[W^{1,\infty}(B_1(O))]^m} \leqq \mu, \tag{6.3.5}$$

have the property

$$\|u\|_{[C^{1,\alpha}(\overline{B_{\frac{1}{2}}(O)})]^m} \leqq c(\mu) \tag{6.3.6}$$

with $\alpha > 0$ (any fixed modulus of continuity plays the same role). Then the condition 6.1.7 is satisfied.

Proof. Let v be a solution to (6.3.4) in $\mathbb{R}^n$ with a bounded gradient. Put $v^R(y) \overset{\text{def}}{=} \frac{1}{R} v(Ry)$, $x = Ry$. The functions $v_R(y)$ are also solutions to (6.3.4) in $\mathbb{R}^n$ and $\|v^R\|_{[W^{1,\infty}(B_1(O))]^m} \leqq \mu$. Hence

$$\left|\frac{\partial v^R}{\partial y_i}(y) - \frac{\partial v^R}{\partial y_i}(0)\right| \leqq c(\mu)\,|y|^\alpha = c(\mu)\left|\frac{x}{R}\right|^\alpha \tag{6.3.7}$$

for $|y| \leqq \frac{1}{2}$ if $R \geqq 2|x|$. Then the passage $R \to \infty$ gives the result, because $\frac{\partial v^R}{\partial y_i}(y) = \frac{\partial v}{\partial x_i}(x) = \frac{\partial v}{\partial x_i}(0)$.

6.3.8. THEOREM. *Let u be a solution to* (6.3.1) *from the family of solutions*

$$\|u\|_{[W^{1,\infty}(\Omega)]^m} + \sum_{r,i} \|f_i^r\|_{W^{1,p}(\Omega)} + \sum_r \|f^r\|_{W^{1,p/2}(\Omega)} \leqq \mu,\ p > n. \tag{6.3.9}$$

Let K be a compact set in Ω. Let (6.3.2) *and the condition $L(\mathbb{R}^n)$ be satisfied. Then*

$$\|u\|_{[C^{1,1-n/p}(K)]^m} \leqq c(\mu, K). \tag{6.3.10}$$

Proof. Let us write $U_s^l \overset{\text{def}}{=} \frac{\partial u_s}{\partial x_l}$. In virtue of Theorem 5.2.11 (in fact, of its trivial modification, because of the condition $u \in [W^{1,\infty}(\Omega)]^m$) we have $u \in [W^{2,2}_{\text{loc}}(\Omega)]^m$. In order to use the results of the preceding section, we turn to the system in variations. Put $u' = \frac{\partial u}{\partial x_l}$. For $\phi \in [\mathscr{D}(\Omega)]^m$ we have, of course, that

$$\int_\Omega \left[a_i^r(x, u, \nabla u)\frac{\partial \phi_r'}{\partial x_i} + a^r(x, u, \nabla u)\,\phi_r' - f_i^r \frac{\partial \phi_r'}{\partial x_i} - f^r \phi_r'\right] dx = 0. \tag{6.3.11}$$

The integration by parts gives

$$(6.3.12)\quad \int_\Omega \left[\frac{\partial a_i^r}{\partial \eta_j^s}\frac{\partial u_s'}{\partial x_i}\frac{\partial \phi_r}{\partial x_i} + \frac{\partial a^r}{\partial \eta_j^s}\frac{\partial u_s'}{\partial x_j}\phi_r\right] dx = \int_\Omega \left[-\frac{\partial a_i^r}{\partial x_l}\frac{\partial \phi_r}{\partial x_i}\right.$$

$$\left. - \frac{\partial a_i^r}{\partial \xi_s} u_s' \frac{\partial \phi_r}{\partial x_i} - \frac{\partial a^r}{\partial x_l}\phi_r - \frac{\partial a^r}{\partial \xi_s} u_s'\phi_r + \frac{\partial f_i^r}{\partial x_l}\frac{\partial \phi_r}{\partial x_i} + \frac{\partial f^r}{\partial x_l}\phi_r\right] dx.$$

Put

$$(6.3.13)\quad \frac{\partial a_i^r}{\partial \eta_j^s}(x, u(x), U) = A_{ij}^{rs}(x, U),$$

$$(6.3.14)\quad \frac{\partial a^r}{\partial \eta_j^s}(x, u(x), U) = A_j^{rs}(x, U).$$

If we take $l = 1, 2, \ldots, n$, we finally get the system

$$(6.3.15)\quad \int_\Omega \left[A_{ij}^{rs}\delta_{kl}\frac{\partial U_s^l}{\partial x_j}\frac{\partial \Phi_r^k}{\partial x_i} + A_j^{rs}\delta_{kl}\frac{\partial U_s^l}{\partial x_j}\Phi_r^k\right] dx$$

$$= \int_\Omega \left[G_i^{rk}\frac{\partial \Phi_r^k}{\partial x_i} + G^{rk}\Phi_r^k\right] dx,$$

where

$$(6.3.16)\quad \sum_{r,k,i} \|G_i^{rk}\|_{L^p(\Omega)} + \sum_{r,k} \|G^{rk}\|_{L^{p/2}(\Omega)} \leqq c_1(\mu).$$

Let us prove that

$$(6.3.17)\quad \lim_{R\to 0} R^{-n} \int_{B_R(x^0)} |U - (U)^{x^0,R}|^2\, dx = 0$$

uniformly with respect to x^0 in K and with respect to all U, generated by u from the family (6.3.9). Let us suppose the contrary. Then there exist $x^\nu \in K$, $x^\nu \to x^0$, u^ν, $R_\nu \to 0$, and $\varepsilon > 0$ such that $R_\nu^{-n} \int_{B_{R_\nu}(x^\nu)} |\nabla u^\nu - (\nabla u^\nu)^{x^\nu, R_\nu}|^2\, dx \geqq \varepsilon^2$.

Put

$$(6.3.18)\quad u^\nu(y) \overset{\text{def}}{=} \frac{1}{R_\nu}[u^\nu(x^\nu + R_\nu y) - u^\nu(x^\nu)].$$

In $\Omega_\nu \to \mathbb{R}^n$ these functions u^ν satisfy the equation

$$(6.3.19)\quad \int_{\Omega_\nu} \left[a_i^r(x^\nu + R_\nu y, R_\nu u^\nu + u^\nu(x^\nu), \nabla_y u^\nu)\frac{\partial \psi_r}{\partial y_i} + R_\nu a^r \psi_r\right] dy$$

$$= \int_{\Omega_\nu} \left[f_i^r(x^\nu + R_\nu y)\frac{\partial \psi_r}{\partial y_i} + R_\nu f^r \psi_r\right] dy.$$

Let $B_a(O) \subset \Omega_\nu$ with a fixed. Put ${}^\nu U_r^l = \dfrac{\partial u_r^\nu}{\partial x_l}$. In Ω_ν, the system in variations has the form

$$(6.3.20)\quad \int_{\Omega_\nu} \left[A_{ij}^{rs}(x^\nu + R_\nu y, R_\nu u^\nu + u^\nu(x^\nu), {}^\nu U)\,\delta_{kl}\frac{\partial\, {}^\nu U_s^l}{\partial y_j}\frac{\partial \Phi_r^k}{\partial y_i}\right.$$

$$\left. + R_\nu A_j^{rs}\delta_{kl}\frac{\partial\, {}^\nu U_s^l}{\partial y_j}\Phi_r^k - R_\nu G_i^{rk}(x^\nu + R_\nu y)\frac{\partial \Phi_r^k}{\partial y_i} - G^{rk}\Phi_r^k\right] dy = 0.$$

Let us look, for example, for

(6.3.21) $$R_\nu^2 \int_{B_2(O)} (G_i^{rk})^2 \, dy = R_\nu^{2-n} \int_{B_{2R_\nu}(x^\nu)} (G_i^{rk})^2 \, dx \leqq c_2 R_\nu^{2\left(1-\frac{n}{p}\right)}.$$

So we get from (6.3.20) that

(6.3.22) $$\|{}^\nu U\|_{[W^{1,2}(B_a(O))]^{mn}} \leqq c_3(a),$$

which implies

(6.3.23) $$\|u^\nu\|_{[W^{2,2}(B_a(O))]^m} \leqq c_4(a).$$

We can thus suppose that $u^\nu \to v$ in $[W^{1,2}(B_a(O))]^m$ $\forall a > 0$. Clearly, v is defined in $\mathbb{R}^n$ and $|\nabla v| \leqq c < \infty$, and it follows from (6.3.19) that v satisfies the system

(6.3.24) $$\int_{\mathbb{R}^n} a_i^r(x^0, \xi, \nabla v) \frac{\partial \phi_r}{\partial x_i} \, dx = 0$$

with $\xi = \lim_{\nu \to \infty} u^\nu(x^\nu)$; we can suppose the existence of such a limit by virtue of (6.3.9). Hence v is a polynomial of at most first degree by $L(\mathbb{R}^n)$. Therefore

(6.3.25) $$\lim_{\nu \to \infty} \int_{B_1(O)} |\nabla u^\nu - (\nabla u^\nu)^{0,1}|^2 \, dy = \int_{B_1(O)} |\nabla v - (\nabla v)^{0,1}|^2 \, dy$$
$$= 0 = \lim_{\nu \to \infty} R_\nu^{-n} \int_{B_{R_\nu}(x^\nu)} |\nabla u^\nu - (\nabla u^\nu)^{x^\nu, R_\nu}|^2 \, dx,$$

which is a contradiction. We can thus apply Lemma 5.2.25 to the system (6.3.15).

Let us study what we can say about the partial regularity of the solution to (6.3.1) under our assumptions. An immediate application of Theorem 6.2.41 is

6.3.26. THEOREM. *Let $u \in [W^{1,\infty}(\Omega)]^M$ be a weak solution to* (6.3.1). *Let* (6.3.2) *be satisfied and let the conditions for the right-hand sides $f_i^r \in W^{1,p}(\Omega)$, $f^r \in W^{1,p/2}(\Omega)$ be satisfied, too. Then there exists $M \subset \Omega$ relatively closed and such that $H_{n-2}(M) = 0$ and $u \in [C^{1,1-n/p}(\Omega \setminus M)]^m$.*

Proof. Indeed, all is reduced to the system (6.3.15) and hence to Theorem 6.2.41 in the case of $U \in [W^{1,2}(\Omega')]^{mn} \cap [L^\infty(\Omega)]^{mn}$, $\bar{\Omega}' \subset \Omega$.

As far as the situation without the condition $u \in [W^{1,\infty}(\Omega)]^m$ is concerned, we shall not go into details because our reduction of the problem to the system (6.3.15) is not convenient in the general case; it is better not to consider quasilinear systems first. We refer the reader to [63] or [74]; in principle, under the conditions of Chapter 5, especially as in Theorem 5.2.69, we get $u \in [C^{1,1-n/p}(\Omega \setminus M)]^m$, $H_{n-2}(M) = 0$, provided that the right-hand sides are regular as in Theorem 6.3.26.

We confine our attention to the special case, when it is obvious how the problem can be transformed into the system (6.3.15). Let us remark, that in Section 6.9. we do not reduce the regularity up to the boundary to the regularity of a solution of a quasilinear system and thus, in fact, the reader can use this process also for the partial regularity.

For the sake of simplicity, let us consider only an L^2 structure. Let us assume the conditions (5.2.8), (5.2.9) and (5.2.10) in $\Omega \times \mathbb{R}^m \times \mathbb{R}^{mn}$. Suppose that the coefficients $\frac{\partial a_i^r}{\partial \eta_j^s}(x, \xi, \eta)$ are uniformly continuous also in $\Omega \times \mathbb{R}^m \times \mathbb{R}^{mn}$.

6.3.27. THEOREM. *Let $u \in [W^{1,2}(\Omega)]^m$ be a weak solution to the system* (6.3.1). *Let $f_i^r \in W^{1,p}(\Omega)$, $f^r \in W^{1,p/2}(\Omega)$, let the coefficients satisfy* (5.2.8)–(5.2.10) *and the uniform continuity of $\frac{\partial a_i^r}{\partial \eta_j^s}$. Let $2 \leqq n \leqq 3$. Then there exists a relatively closed set M in Ω, such that $H_{n-2}(M) = 0$ and $u \in [C^{1,1-n/p}(\Omega \setminus M)]^m$. (For $n = 2$ it implies that $M = \emptyset$.)*

Proof. Once more, all is reduced to the system (6.3.15) and to Theorem 6.2.41, its second case, because $W^{2,2}(\Omega') \subsetneq C^{0,2/3}(\bar{\Omega}')$ for $n = 3$. $\bar{\Omega}' \subset \Omega$, and $W^{2,2}(\Omega') \subsetneq C^{0,\mu}(\bar{\Omega}')$ for $n = 2$, $\mu > 0$.

6.4. A counter-example

We shall reproduce here the counter-example from the works [16], [21], [70]. Let us begin with the observation that the counter-example of E. Giusti, M. Miranda [61] mentioned in 6.1. as (6.1.10) is not the system in variations (6.3.15). So we shall look for u such that $|\nabla u| \leqq c < \infty$ in $\mathbb{R}^n$, that

(6.4.1) $$-\frac{\partial}{\partial x_i}[a_i^r(\nabla u)] = 0,$$

provided that $a_i^r(\eta)$ are real analytic functions, that the condition of very strong ellipticity is satisfied, and that u is not a polynomial of at most first degree.

6.4.2. THEOREM. *Let $u_{ij} \overset{\text{def}}{=} \frac{x_i x_j}{|x|} - \frac{1}{n}\delta_{ij}|x|$. Further let $\nabla_i u = \frac{\partial u_{ij}}{\partial x_j}$, $\delta_i u = \frac{\partial u_{jj}}{\partial x_i}$, $\nabla_{ijk} u = \frac{\partial u_{ij}}{\partial x_k} + \gamma[\nabla_i u \delta_{jk} + \nabla_j u \delta_{ik} + \nabla_k u \delta_{ij}]$. Put $\gamma = 0$ for $n \geqq 4$, let $-\frac{4}{15} < \gamma < \frac{-4 + \sqrt{14}}{12}$ for $n = 3$. Let $V = \bar{\nu}\left(n - \frac{1}{n}\right)^5$. Further let $A(n) = n + 1 + 4\gamma\left(n - \frac{1}{n}\right) + 3\gamma^2(n + 2)\left(n - \frac{1}{n}\right)$, $B(n) = \left(n - \frac{1}{n}\right) \times$ $\times (n - 1 - 3\gamma)^2$ and let V_1, V_2 be the roots of the equation*

(6.4.3) $$-\frac{3}{4B(n)}V^2 + V\left[1 - \frac{3}{2}\frac{A(n)}{B(n)}\right] - \frac{3}{4}\frac{A(n)^2}{B(n)} = 0.$$

Let $V_1 < V < V_2$. Then $V_1 > 0$ and if we put

(6.4.4) $$\lambda = \frac{n + 1 + 4\gamma(n - n^{-1}) + 3\gamma^2(n + 2)(n - n^{-1}) + \bar{\nu}(n - n^{-1})^5}{(n - 1)(n - n^{-1})^3 - 3\gamma(n - n^{-1})^3},$$

$$\delta = -\frac{1 + n^{-1} + 4\gamma(n - n^{-1}) + (n - 1)(n - n^{-1})}{(n - 1)(n - n^{-1})} - \frac{3\lambda\gamma(n - n^{-1})^3 + 3\gamma^2 n(n - n^{-1}) + 6\gamma^2(n - n^{-1}) + V}{(n - 1)(n - n^{-1})}, \tag{6.4.5}$$

and take β large enough, then in the space of symmetric matrices $u_{ij} = u_{ji}$, with $|\nabla u_{ij}| \leqq c < \infty$, we get the counter-example in the form

$$\int_{\mathbb{R}^n} \nabla_{ijk}u \nabla_{ijk}\phi \, dx + \delta \int_{\mathbb{R}^n} \nabla_i u \delta_i \phi \, dx + \beta \int_{\mathbb{R}^n} \delta_i u \delta_i \phi \, dx + \lambda \int_{\mathbb{R}^n} \nabla_i u \nabla_j u \nabla_k u \nabla_{ijk}\phi \, dx + \bar{v} \int_{\mathbb{R}^n} (\nabla_i u \nabla_i u)^2 \nabla_j u \nabla_j \phi \, dx = 0. \tag{6.4.6}$$

Let us once more remark that the system (6.4.6) is not Euler's one.

Let us also remark that if $u(x_1, x_2, x_3)$ is the counter-example for $n = 3$, then define for $n \geqq 4$: $u(x_1, x_2, x_3, \ldots, x_n) \overset{\text{def}}{=} u(x_1, x_2, x_3)$. Then the set M of the singular points for such u has $\operatorname{Dim} M = 0$. If $n = 3$, and we take $\gamma \in \left(-\frac{4}{15}, \frac{-4 + \sqrt{14}}{12} \right)$ and $V \in (V_1(\gamma), V_2(\gamma))$, then at least two-dimen-sional manifolds lie in the space of the operators without $L(\mathbb{R}^n)$.

P r o o f of Theorem 6.4.2. First we get that u_{ij} is a weak solution to (6.4.6). Let us substitute $\tilde{\eta}_k^{ij}, \tilde{v}_i, \tilde{\mu}_i, \tilde{\zeta}_{ijk}$ for $\frac{\partial \phi_{ij}}{\partial x_k}$, $\nabla_i \phi$, $\delta_i \phi$, $\nabla_{ijk}\phi$; and $\eta_k^{ij}, v_i, \mu_i, \zeta_{ijk}$ for $\frac{\partial u_{ij}}{\partial x_k}$, $\nabla_i u$, $\delta_i u$, $\nabla_{ijk} u$. We have

$$\frac{\partial a_k^{ij}}{\partial \eta_\delta^{\alpha\beta}} \tilde{\eta}_k^{ij} \tilde{\eta}_\delta^{\alpha\beta} = |\tilde{\zeta}|^2 + \delta \tilde{v}_i \tilde{\mu}_i + \beta \tilde{\mu}_i \tilde{\mu}_i + \lambda(\tilde{v}_i v_j v_k + v_i \tilde{v}_j v_k + v_i v_j \tilde{v}_k)\tilde{\zeta}_{ijk} + 4\bar{v}|v|^2 (v_i \tilde{v}_i)^2 + \bar{v}|v|^4 |\tilde{v}|^2 \geqq |\tilde{\zeta}|^2 + \bar{v}[2(v, \tilde{v})^2 |v|^2 + |\tilde{v}|^2 |v|^4] - |\lambda| |\tilde{\zeta}| [2(v, \tilde{v})^2 |v|^2 + |\tilde{v}|^2 |v|^4]^{\frac{1}{2}} \sqrt{3} + \beta|\tilde{\mu}|^2 - \delta|\tilde{v}| |\tilde{\mu}|. \tag{6.4.7}$$

Let us suppose for a moment that we have proved, putting $\tilde{\varkappa}^2 = 2(v, \tilde{v})^2 |v|^2 + |\tilde{v}|^2 |v|^4$, that

$$|\tilde{\zeta}|^2 + \bar{v}\tilde{\varkappa}^2 - |\lambda| \sqrt{3}|\tilde{\zeta}| \tilde{\varkappa} \geqq c_1|\tilde{\zeta}|^2, \quad c_1 > 0. \tag{6.4.8}$$

We have $|\tilde{\zeta}|^2 = |\tilde{\eta}|^2 + 4\gamma|\tilde{v}|^2 + 2\gamma(\tilde{v}, \tilde{\mu}) + 3\gamma^2(n + 2)|\tilde{v}|^2$. Let us prove:

$$|\tilde{\eta}|^2 + 4\gamma|\tilde{v}|^2 + 3\gamma^2(n + 2)|\tilde{v}|^2 \geqq c_2|\tilde{\eta}|^2. \tag{6.4.9}$$

But $|\tilde{v}| \leqq \sqrt{n}|\tilde{\eta}|$, hence $|\tilde{\eta}|^2 + 4\gamma|\tilde{v}|^2 + 3\gamma^2(n + 2)|\tilde{v}|^2 \geqq |\tilde{\eta}|^2[1 + 4\gamma n + 3n(n + 2)\gamma^2]$, because $\gamma \leqq 0$ and $\gamma \geqq -\frac{4}{3(n + 2)}$. It is thus sufficient that $1 + 4\gamma n + 3n(n + 2)\gamma^2 > 0$ but this is true for $\gamma = 0$ and for $n = 3$ also because $1 + 12\gamma + 45\gamma^2 > 0$. So if we have assumed (6.4.8), we get

$$|\tilde{\zeta}|^2 + \bar{v}\tilde{\varkappa}^2 - \sqrt{3}|\lambda| |\tilde{\zeta}| \tilde{\varkappa} + \beta|\tilde{\mu}|^2 - |\delta| |\tilde{v}| |\tilde{\mu}| \geqq c_3|\tilde{\eta}|^2, \tag{6.4.10}$$

because we can choose β as large as we wish: we have $\delta_i u = 0$ for our solu-tion.

But (6.4.8) follows from the inequality

(6.4.11) $$\bar{\nu} - \tfrac{3}{4}\lambda^2 > 0$$

and (6.4.11) is true if (6.4.3) has one positive root. But if

(6.4.12) $$1 - 3\frac{A(n)}{B(n)} > 0,$$

both the roots of (6.4.3) are positive. The inequality (6.4.12) is equivalent to the inequality

(6.4.13) $$9(n+1)(n-1)\gamma^2 + 6\gamma(n-1)(n+1) + 3n - (n-1)^3 < 0.$$

For $n \geqq 4$ and $\gamma = 0$, the inequality (6.4.13) is satisfied. For $n = 3$ it is true, if we take $\gamma \in \left(\frac{-4-\sqrt{14}}{12}, \frac{-4+\sqrt{14}}{12}\right)$; but $-\frac{4}{15} < \frac{-4+\sqrt{14}}{12}$; we can thus satisfy also the inequality $\gamma \geqq -\frac{4}{15}$.

6.5. The dimension $n = 2$, one equation of the second order, Saint-Venant's principle

6.5.1. THEOREM. *If $n = 2$ then the condition $L(\mathbb{R}^n)$ is satisfied.*

In fact, Theorem 6.5.1 is clear, because the interior regularity for $n = 2$ is proved, for example, in [13] and by 6.3.3, $L(\mathbb{R}^n)$ follows. Let us further mention that the regularity, in the case $n = 2$, follows also from 6.3.27.

Proof of Theorem 6.5.1: Let $x^0 \in \Omega$, $\xi \in \mathbb{R}^m$, v be a solution to

(6.5.2) $$\int_{\mathbb{R}^2} a_i^r(x^0, \xi, \nabla v)\frac{\partial \phi_r}{\partial x_i}\,dx = 0$$

with $|\nabla v| \leqq c < \infty$. In virtue of Theorem 5.2.11 we have $v \in [W^{2,2}_{\mathrm{loc}}(\mathbb{R}^2)]^m$ hence we get

(6.5.3) $$\int_{\mathbb{R}^2} \frac{\partial a_i^r}{\partial \eta_j^s}(x^0, \xi, \nabla v)\frac{\partial v_s'}{\partial x_j}\frac{\partial \phi_r}{\partial x_i}\,dx = 0$$

for $v' = \dfrac{\partial v}{\partial x_l}$. Let $\eta \in \mathscr{D}(B_{2T}(O))$, $\eta(x) = 1$, for $|x| \leqq T$, $0 \leqq \eta \leqq 1$, $|\nabla \eta| \leqq \dfrac{c_1}{T}$. Put $\phi = v'\eta^2$. Then we have

$$0 = \int_{\mathbb{R}^2} \frac{\partial a_i^r}{\partial \eta_j^s}\frac{\partial v_s'}{\partial x_j}\frac{\partial v_r'}{\partial x_i}\eta^2\,dx + \int_{\mathbb{R}^2} \frac{\partial a_i^r}{\partial \eta_j^s}\frac{\partial v_s'}{\partial x_j}2\eta\frac{\partial \eta}{\partial x_i}v_r\,dx,$$

which implies

(6.5.4) $$\left(\int_{\mathbb{R}^2}(|\nabla v'|^2\eta^2\,dx\right)^{\frac{1}{2}} \leqq c_2\|\nabla v\|_{[L^\infty(\mathbb{R}^n)]^{2m}} \leqq c_3 < \infty.$$

The inequality (6.5.4) implies that $\int_{\mathbb{R}^2} |\nabla v'|^2 \, dx < \infty$. Now, there exist $\phi_k \in \mathscr{D}(\mathbb{R}^2)$ such that $\nabla \phi_k \to \nabla v'$ in $[L^2(\mathbb{R}^2)]^{2m}$, therefore (6.5.3) implies that $\int_{\mathbb{R}^2} \frac{\partial a_i^r}{\partial \eta^s} \frac{\partial v_s'}{\partial x_j} \frac{\partial v_r'}{\partial x_i} \, dx = 0$.

6.5.5. REMARK. If we repeat the proof of Theorem 6.5.1 for $\mathbb{R}^n$, we get for a solution to (6.5.2) with $|\nabla v| \leqq c < \infty$ in $\mathbb{R}^n$ that

$$\int_{B_T(O)} |\nabla v'|^2 \, dx \leqq c_1 \|\nabla v\|_{[L^\infty(\mathbb{R}^n)]^{nm}} \cdot T^{n-2}.$$

6.5.6. THEOREM. *If $m = 1$ then the condition $L(\mathbb{R}^n)$ is satisfied.*

Of course, this follows from the regularity, see, for example, Theorem 5.4.32, and from Theorem 6.3.3. The "more direct" proof: We consider (6.5.2) ($m = 1$) and the equation in variations (6.5.3). Then we use 5.4.20.

As in Section 6.3., let us consider quasilinear systems (of a special form)

$$\text{(6.5.7)} \qquad -\frac{\partial}{\partial x_i}\left[A_{ij}^{rs}(x, v) \frac{\partial v_s}{\partial x_j}\right] = 0$$

provided that $v \in [W^{1,2}(B_\varrho(O))]^M \cap [L^\infty(B_\varrho(O))]^M$. Let $\|v\|_{[L^\infty(B_\varrho(O))]^M} \leqq \mu$, $|\eta| \leqq \mu$ (in $\mathbb{R}^M$) and let $\forall x^0 \in B_\varrho(O)$, $\forall |\eta| \leqq \mu$,

$$\text{(6.5.8)} \qquad \lambda_0(\mu) \, |\xi|^2 \leqq A_{ij}^{rs}(x^0, \eta) \, \xi_i^r \xi_j^s \leqq \lambda_1(\mu) \, |\xi|^2, \quad \lambda_0(\mu) > 0.$$

Put also

$$\text{(6.5.9)} \qquad A_{ij}^{rs}(x^0, \eta) \, \xi_i^r \zeta_j^s \leqq \lambda_2(\mu) \, (A_{ij}^{rs}(x^0, \eta) \, \xi_i^r \xi_j^s)^{\frac{1}{2}} \, (A_{ij}^{rs}(x^0, \eta) \, \zeta_i^r \zeta_j^s)^{\frac{1}{2}}.$$

We shall now follow an idea of O. A. Oleĭnik, see [69].

6.5.10. THEOREM. *Let v with $\|v\|_{[L^\infty(B_\varrho(O))]^M} \leqq \mu$ be a solution to (6.5.7). Let $0 \leqq \varrho_0 \leqq \varrho_1 \leqq \varrho$. Further let $\gamma = \left[\frac{\lambda_0(\mu)}{\lambda_1(\mu)}\right]^{\frac{1}{2}} [\lambda_2(\mu)]^{-1} (n-1)^{\frac{1}{2}}$. Then*

$$\text{(6.5.11)} \qquad \int_{B_{\varrho_0}} A_{ij}^{rs}(x, v) \frac{\partial v_r}{\partial x_i} \frac{\partial v_s}{\partial x_j} \, dx \leqq \left(\frac{\varrho_0}{\varrho_1}\right)^\gamma \int_{B_{\varrho_1}} A_{ij}^{rs}(x, v) \frac{\partial v_r}{\partial x_i} \frac{\partial v_s}{\partial x_j} \, dx.$$

Proof: Put $I(\tilde{\varrho}) = \int_{B_{\tilde{\varrho}}(O)} A_{ij}^{rs} \frac{\partial v_r}{\partial x_i} \frac{\partial v_s}{\partial x_j} \, dx = \int_0^{\tilde{\varrho}} r^{n-1} \int_{\partial B_1(O)} L(r, \xi) \, dS$. Thus $I'(\tilde{\varrho}) = \int_{\partial B_{\tilde{\varrho}}(O)} L(r, \xi) \, dS$ almost everywhere in $(0, \varrho)$. Let $U(\tilde{\varrho}) = \int_{B_{\tilde{\varrho}}(O)} |v|^2 \, dx$, thus $U'(\tilde{\varrho}) = \int_{\partial B_{\tilde{\varrho}}(O)} |v|^2 \, dS$ almost everywhere in $(0, \varrho)$. Now

$$\text{(6.5.12)} \qquad \int_0^\varrho I(\tilde{\varrho}) \, d\tilde{\varrho} = \int_{B_\varrho(O)} L(r, \xi) \, (\varrho - |x|) \, dx,$$

hence putting $\phi = (v - (v)^{\varrho'})(\varrho - |x|)$, $0 \leqq \varrho' \leqq \varrho$, where $(v)^{\varrho'} = \frac{1}{|\partial B_{\varrho'}(O)|} \int_{\partial B_{\varrho'}(O)} v \, dS$, we have

$$(6.5.13) \qquad 0 = \int_{B_\varrho(O)} A_{ij}^{rs} \frac{\partial v_s}{\partial x_j} \frac{\partial \phi_r}{\partial x_i} \, dx = \int_0^\varrho I(\tilde\varrho) \, d\tilde\varrho - \int_{B_\varrho(O)} A_{ij}^{rs} \frac{\partial v_s}{\partial x_j} (v_r - (v_r)^{\varrho'}) \frac{x_i}{|x|} \, dx.$$

Therefore it follows from (6.5.13) that

$$(6.5.14) \qquad I(\tilde\varrho) = \int_{\partial B_{\tilde\varrho}(O)} A_{ij}^{rs} \frac{\partial v_s}{\partial x_j} (v_r - (v_r)^{\varrho'}) \nu_i \, dS$$

almost everywhere in $(0, \varrho)$, thus (6.5.14) is valid almost everywhere in $(0, \varrho)$ if $\tilde\varrho = \varrho'$.

Actually,

$$(6.5.15) \qquad I(\tilde\varrho) \leqq \lambda_2 [I'(\tilde\varrho)]^{\frac{1}{2}} \lambda_1^{\frac{1}{2}} \Big(\int_{\partial B_{\tilde\varrho}(O)} |v - (v)^{\tilde\varrho}|^2 \, dS \Big)^{\frac{1}{2}}.$$

It follows from the properties of Laplace-Beltrami operator that

$$(6.5.16) \qquad \int_{\partial B_{\tilde\varrho}(O)} |v - (v)^{\tilde\varrho}|^2 \, dS \leqq \tilde\varrho^2 (n-1)^{-1} \int_{\partial B_{\tilde\varrho}(O)} |\nabla v|^2 \, dS.$$

If we combine (6.5.15) and (6.5.16) we get

$$I(\tilde\varrho) \leqq \lambda_2 \left(\frac{\lambda_1}{\lambda_0}\right)^{\frac{1}{2}} \tilde\varrho (n-1)^{-\frac{1}{2}} I'(\tilde\varrho),$$

which is equivalent to $\frac{I'(\tilde\varrho)}{I(\tilde\varrho)} \geqq \lambda_2^{-1} \left(\frac{\lambda_0}{\lambda_1}\right)^{\frac{1}{2}} (n-1)^{\frac{1}{2}} \frac{1}{\tilde\varrho}$. The integration of the last inequality over (ϱ_0, ϱ_1) gives the result.

6.5.17. THEOREM. *Let* $v \in [L^\infty(R^n)]^M$, $\|v\|_{[L^\infty(\mathbb{R}^n)]^M} \leqq \mu$, $v \in [W^{1,2}_{\mathrm{loc}}(\mathbb{R}^n)]^M$ *be a weak solution to the system* (6.5.7). *Let* (6.5.8) *and* (6.5.9) *be valid* $\forall x^0 \in \mathbb{R}^n$. *Let* $2 \leqq n \leqq 3$. *Then* $v = \text{const}$ *if* $(n-1)^{\frac{1}{2}} \lambda_2^{-1} \left(\frac{\lambda_0}{\lambda_1}\right)^{\frac{1}{2}} > n - 2$.

Proof. In virtue of Remark 6.5.5 it is $\int_{B_\varrho(O)} |\nabla v|^2 \, dx \leqq c\varrho^{n-2}$. It follows from (6.5.11) that $\int_{B_{\varrho_0}(O)} |\nabla v|^2 \, dx \leqq c_1 \left(\frac{\varrho_0}{\varrho_1}\right)^\gamma \int_{B_{\varrho_1}(O)} |\nabla v|^2 \, dx \leqq c_2 \varrho_0^\gamma \varrho_1^{-\gamma+n-2}$. If $\varrho_1 \to \infty$, then $\int_{B_{\varrho_0}(O)} |\nabla v|^2 \, dx = 0$, $\forall \varrho_0$.

For $u \in W^{1,2}(B_r(O))$, put $|\partial_\Theta u|^2 \overset{\text{def}}{=} r^2 \left[|\nabla u|^2 - \left(\frac{\partial u}{\partial r}\right)^2 \right]$.

6.5.18. EXERCISE. Prove: if $u \in W^{1,2}(B_r(O))$, $\Delta u = 0$ in $B_r(O)$ and $\int_{\partial B_r} |\partial_\Theta u|^2 \, dS < \infty$, then

$$\int_{B_r(O)} |\nabla u|^2 \, dx \leqq (n-1)^{-1} r^{-1} \int_{\partial B_r(O)} |\partial_\theta u|^2 \, dS.$$

6.5.19. THEOREM. *Let us consider a system* (6.5.7) *such that* $A_{ij}^{rs}(x, \eta) = A_{ji}^{sr}(x, \eta)$. *Let* $0 \leqq \varrho_0 \leqq \varrho_1 \leqq \varrho$ *and* v *be a weak solution to* (6.5.7) *with* $\|v\|_{[L^\infty(B_\varrho(O))]^M} \leqq \mu$. *Then* (6.5.11) *is true with* $\gamma \geqq \dfrac{\lambda_0}{\lambda_1}(n-1)$.

Proof. Note that $(Au, u) \overset{\text{def}}{=} A_{ij}^{rs}(x, v) \dfrac{\partial u_r}{\partial x_i} \dfrac{\partial u_s}{\partial x_j}$. Let $h(\tilde{\varrho}) = \tilde{\varrho}^{-\gamma} \displaystyle\int_{B_{\tilde{\varrho}}(O)} (Av, v) \, dx$.

Almost everywhere in $(0, \varrho)$ we have $h'(\tilde{\varrho}) = -\gamma \tilde{\varrho}^{-\gamma-1} \int_{B_{\tilde{\varrho}}(O)} (Av, v) \, dx + \tilde{\varrho}^{-\gamma} \int_{\partial B_{\tilde{\varrho}}(O)} (Au, u) \, dS$, hence $h'(\tilde{\varrho}) \geqq 0 \Leftrightarrow$

$$\gamma \int_{B_{\tilde{\varrho}}(O)} (Au, u) \, dx \leqq \tilde{\varrho} \int_{\partial B_{\tilde{\varrho}}(O)} (Au, u) \, dS. \tag{6.5.20}$$

But $\tilde{\varrho} \displaystyle\int_{\partial B_{\tilde{\varrho}}(O)} (Au, u) \, dS \geqq \lambda_0 \tilde{\varrho} \int_{\partial B_{\tilde{\varrho}}(O)} |\nabla u|^2 \, dS \geqq \lambda_0 \tilde{\varrho} \int_{\partial B_{\tilde{\varrho}}(O)} \frac{1}{\tilde{\varrho}^2} |\partial_\vartheta u|^2 \, dS \geqq \lambda_0 (n-1) \int_{B_{\tilde{\varrho}}(O)} |\nabla \tilde{u}|^2 \, dx \geqq \frac{\lambda_0}{\lambda_1}(n-1) \int_{B_{\tilde{\varrho}}(O)} (A\tilde{u}, \tilde{u}) \, dx \geqq \frac{\lambda_0}{\lambda_1}(n-1) \int_{B_{\tilde{\varrho}}(O)} (Au, u) \, dx$; here $\tilde{u}$ is the solution to $\Delta \tilde{u} = 0$ in $B_{\tilde{\varrho}}(O)$ with $\tilde{u} = u$ on $\partial B_{\tilde{\varrho}}(O)$.

6.5.21. THEOREM. *Let* v *be a weak solution to the system* (6.5.7) *with* $\|v\|_{[L^\infty(\mathbb{R}^n)]^M} \leqq \mu$ *in* $\mathbb{R}^n$. *Let* (6.5.8) *be valid for* $\forall x^0 \in \mathbb{R}^n$. *Then* $v = \text{const}$ *if* $\dfrac{\lambda_0}{\lambda_1} > \dfrac{n-2}{n-1}$.

Proof. We proceed as in the proof of Theorem 6.5.17, using Remark 6.5.5 and Theorem 6.5.19.

6.5.22. REMARK. As we have just mentioned, Theorem 6.5.21 is true for

$$\frac{\lambda_0}{\lambda_1} > \frac{\sqrt{1 + \dfrac{(n-2)^2}{n-1}} - 1}{\sqrt{1 + \dfrac{(n-2)^2}{n-1}} + 1}, \text{ see [21].}$$

6.6. Regularity up to the boundary via the fourth order differential

For details, see the work [71]. Here, we shall be interested only in the case $\Omega = \mathbb{R}^3_+$. We suppose that our system is Euler's one. So, let us consider the scalar product in $[W^{3,2}(\mathbb{R}^3_+)]^m$ in the form

(6.6.1) $$\int_{\mathbb{R}^3_+} \sum_{1 \leqq |\alpha| \leqq 3} D^\alpha u_r D^\alpha v_r \, dx$$

and let us denote the corresponding seminorm by

(6.6.2) $$\|u\|'_3.$$

In $W^{3,2}(\mathbb{R}^3_+)$, we consider the norm in the form

(6.6.3) $$\|u\|_3^2 \overset{\text{def}}{=} \|u\|'^2_3 + \int_{B_1(O) \cap \mathbb{R}^3_+} |u|^2 \, dx.$$

Let C be the space of constant functions in $\mathbb{R}^3_+$.

6.6.4. LEMMA. *In the space* $[W^{3,2}(\mathbb{R}^3_+)]^m$, *the norm* (6.6.2) *is an equivalent norm.*

Proof is clear from Poincaré's inequality

(6.6.5) $$\int_{B_1(O) \cap \mathbb{R}^3_+} \left| u - \frac{1}{|B_1(O) \cap \mathbb{R}^3_+|} \int_{B_1(O) \cap \mathbb{R}^3_+} u(y) \, dy \right|^2 dx \leqq c_1 \int_{B_1(O) \cap \mathbb{R}^3_+} |\nabla u|^2 \, dx.$$

Let us write $V = \{v \in [W^{3,2}(\mathbb{R}^3)]^m \mid v = 0 \text{ on } \partial\mathbb{R}^3_+\}$ and $\tilde{V} = (V + C)/C$. Let

$$\|u\|'_{1,\infty} \overset{\text{def}}{=} \left\| \left(\frac{\partial u_r}{\partial x_i} \frac{\partial u_r}{\partial x_i} \right)^{\frac{1}{2}} \right\|_{L^\infty(\mathbb{R}^n_+)}.$$

Let now $0 < a < \infty$ and F be a function from $C^4\big(\overline{(B_{a+\varepsilon}(O)}\big)$, where $B_{a+\varepsilon}(O) \subset \mathbb{R}^{3m}$. For $|\eta| \leqq a + \varepsilon$ let

(6.6.6) $$\frac{\partial^2 F}{\partial \eta_i^r \, \partial \eta_j^s}(\eta) \, \zeta_i^r \zeta_j^s \geqq \nu |\zeta|^2, \quad \nu > 0,$$

(6.6.7) $$\frac{\partial^4 F}{\partial \eta_i^r \, \partial \eta_j^s \, \partial \eta_k^t \, \partial \eta_l^v}(\eta) \, \zeta_i^r \zeta_j^s \zeta_k^t \zeta_l^v \leqq T |\zeta|^4,$$

(6.6.8) $$\nu - 3a^2 T > 0.$$

Let also $f \in [W^{2,2}(\mathbb{R}^3_+)]^{3m}$, where the space $[W^{2,2}(\mathbb{R}^3_+)]^{3m}$ is endowed with the usual scalar product $\int_{\mathbb{R}^3_+} \sum_{|\alpha| \leqq 2} D^\alpha f_i^r D^\alpha g_i^r \, dx$. Let $\tilde{u}^0 \in [W^{3,2}(\mathbb{R}^3_+)]^{3m}/C \overset{\text{def}}{=} \tilde{W}$; $u \in \tilde{u}$ is understood to be a representative.

6.6.9. DEFINITION. *We look for* $\tilde{u} \in \tilde{W}$ *such that* $\tilde{u} - \tilde{u}^0 \in \tilde{V}$ *and that* $\forall \tilde{\phi} \in \tilde{V}$ *we have*

$$\int_{\mathbb{R}^3_+} \left[\frac{\partial F}{\partial \eta_i^r}(\nabla u) \frac{\partial \phi_r}{\partial x_i} - f_i^r \frac{\partial \phi_r}{\partial x_i} \right] dx = 0, \quad \|u\|'_{1,\infty} \leqq a.$$

6.6.10. LEMMA. *The space* $W^{2,2}(\mathbb{R}^3_+)$ *is a Banach algebra.*

Proof. Let $f \in W^{2,2}(\mathbb{R}^3_+)$ and extend it to $\mathbb{R}^3_-$ in such a way that

(6.6.11) $$\|f\|_{W^{2,2}(\mathbb{R}^3)} \leqq c \|f\|_{W^{2,2}(\mathbb{R}^3_+)}.$$

(See, for example, the NIKOLSKIĬ-BABIČ extension in [25].) For the Fourier transform $\hat{f}$ of f we have

$$(6.6.12)\qquad c_1 \int_{\mathbb{R}^3} |\hat{f}|^2(1 + |\xi|^2)^2 \, d\xi \leqq \|f\|^2_{W^{2,2}(\mathbb{R}^3)} \leqq c_2 \int_{\mathbb{R}^3} |\hat{f}|^2 (1 + |\xi|^2)^2 \, d\xi .$$

Hence it follows from the inverse transformation

$$(6.6.13)\qquad f(x) = \frac{1}{(2\pi)^3} \int_{\mathbb{R}^3} \hat{f}(\xi)\, e^{i(x,\xi)} \, d\xi$$

that

$$(6.6.14)\qquad \|f\|_{L^\infty(\mathbb{R}^3)} \leqq c_3 \|f\|_{W^{2,2}(\mathbb{R}^3)} .$$

We also see that $\overline{\mathscr{D}(\mathbb{R}^3)\,|_{\mathbb{R}^3_+}} = W^{2,2}(\mathbb{R}^3_+)$. We have the following interpolation inequality (a special case of the Nirenberg-Gagliardo inequality): Let $\Psi \in \mathscr{D}(\mathbb{R}^1)$, then

$$\int_{-\infty}^{\infty} \Psi'^4 \, dx = -3 \int_{-\infty}^{\infty} \Psi \Psi'^2 \Psi'' \, dx \leqq 3 \max_{x \in \mathbb{R}^1} |\Psi(x)| \left(\int_{-\infty}^{\infty} \Psi'^4 \, dx \right)^{\frac{1}{2}} \left(\int_{-\infty}^{\infty} \Psi''^2 \, dx \right)^{\frac{1}{2}},$$

and from this we finally obtain

$$(6.6.15)\qquad \int_{-\infty}^{\infty} \Psi'^4 \, dx \leqq 9 \max_{x \in \mathbb{R}^1} \Psi^2(x) \int_{-\infty}^{\infty} \Psi''^2 \, dx .$$

So if $f, g \in W^{2,2}(\mathbb{R}^3_+)$, then

$$\|fg\|_{L^2(\mathbb{R}^3_+)} \leqq \|f\|_{L^\infty(\mathbb{R}^3_+)} \|g\|_{L^2(\mathbb{R}^3_+)},$$

$$\left\| \frac{\partial f}{\partial x_i} g \right\|_{L^2(\mathbb{R}^3_+)} \leqq \|g\|_{L^\infty(\mathbb{R}^3_+)} \|f\|_{W^{1,2}(\mathbb{R}^3_+)},$$

$$\left\| \frac{\partial^2 f}{\partial x_i \, \partial x_j} g \right\|_{L^2(\mathbb{R}^3_+)} \leqq \|g\|_{L^\infty(\mathbb{R}^3_+)} \|f\|_{W^{2,2}(\mathbb{R}^3_+)},$$

$$\left\| \frac{\partial f}{\partial x_i} \frac{\partial g}{\partial x_j} \right\|_{L^2(\mathbb{R}^3_+)} \leqq \left\| \frac{\partial f}{\partial x_i} \right\|_{L^4(\mathbb{R}^3_+)} \left\| \frac{\partial g}{\partial x_j} \right\|_{L^4(\mathbb{R}^3_+)}$$
$$\leqq 3 \|f\|^{\frac{1}{2}}_{L^\infty(\mathbb{R}^3_+)} \|f\|^{\frac{1}{2}}_{W^{2,2}(\mathbb{R}^3_+)} \|g\|^{\frac{1}{2}}_{L^\infty(\mathbb{R}^3_+)} \|g\|^{\frac{1}{2}}_{W^{2,2}(\mathbb{R}^3_+)} .$$

6.6.16. THEOREM. *Let u be a solution to (6.6.9) and let* $\dfrac{\partial F}{\partial \eta_i^r}(0) = 0$, $r = 1, 2, \ldots, m$, $i = 1, 2, 3$. *Then*

$$(6.6.17)\qquad \|u\|'_3 \leqq c[\|f\|_2 + \|f\|_2^2 + \|u^0\|'_3] .$$

Proof. Let $\phi \in [\mathscr{D}(\mathbb{R}^3_+)]^m$, $\phi' = \dfrac{\partial \phi}{\partial x_l}$, $l = 1$ or 2. Then

$$(6.6.18)\qquad \int_{\mathbb{R}^3_+} f_i^{r\prime\prime} \frac{\partial \phi_r''}{\partial x_i} \, dx = \int_{\mathbb{R}^3_+} \frac{\partial^2 F}{\partial \eta_i^r \, \partial \eta_j^s} \frac{\partial u_s''}{\partial x_j} \frac{\partial \phi_r''}{\partial x_i} \, dx$$
$$+ \int_{\mathbb{R}^3_+} \frac{\partial^3 F}{\partial \eta_i^r \, \partial \eta_j^s \, \partial \eta_k^t} \frac{\partial u_s'}{\partial x_j} \frac{\partial \phi_r''}{\partial x_i} \frac{\partial u_t'}{\partial x_k} \, dx .$$

We can actually substitute $u - u^0$ for ϕ in (6.6.18). We first have

$$\int_{\mathbb{R}^3_+} \frac{\partial^3 F}{\partial\eta_i^r \partial\eta_j^s \partial\eta_k^t} \frac{\partial u_s'}{\partial x_j} \frac{\partial u_r''}{\partial x_i} \frac{\partial u_t'}{\partial x_k} \,\mathrm{d}x = -\frac{1}{3}\int_{\mathbb{R}^3_+} \frac{\partial^4 F}{\partial\eta_i^r \partial\eta_j^s \partial\eta_k^t \partial\eta_l^v} \frac{\partial u_r'}{\partial x_i} \frac{\partial u_s'}{\partial x_j} \frac{\partial u_t'}{\partial x_k} \frac{\partial u_v'}{\partial x_l} \,\mathrm{d}x. \tag{6.6.19}$$

Now, if we use (6.6.15) we get a bound for the terms on the right-hand side with u^0 in the form

$$c_1 \|u''\|_1' \|u^{0''}\|_1'. \tag{6.6.20}$$

But it follows from (6.6.15) that (because $|\nabla u| \leqq a$)

$$\int_{\mathbb{R}^3_+} \left[\sum_{i,j} \left(\frac{\partial u_i'}{\partial x_j}\right)^2\right]^2 \mathrm{d}x \leqq 9a^2 \int_{\mathbb{R}^3_+} \sum_{i,j} \left(\frac{\partial u_i''}{\partial x_j}\right)^2 \mathrm{d}x, \tag{6.6.20'}$$

hence we get from (6.6.19) and (6.6.20) that

$$-\frac{1}{3}\int_{\mathbb{R}^3_+} \frac{\partial^4 F}{\partial\eta_i^r \partial\eta_j^s \partial\eta_k^t \partial\eta_l^v} \frac{\partial u_r'}{\partial x_i} \frac{\partial u_s'}{\partial x_j} \frac{\partial u_t'}{\partial x_k} \frac{\partial u_v'}{\partial x_l} \,\mathrm{d}x \geqq -3Ta^2 \int_{\mathbb{R}^3_+} \sum_{r,i} \left(\frac{\partial u_r''}{\partial x_i}\right)^2 \mathrm{d}x. \tag{6.6.21}$$

We thus obtain from (6.6.18) that

$$\|u''\|_1'^2 \leqq c_2[\|f''\|_0^2 + \|u^{0''}\|_1'^2]. \tag{6.6.22}$$

It follows from (6.6.15) and (6.6.22) that

$$\int_{\mathbb{R}^3_+} \sum_{r,i} \left(\frac{\partial u_r'}{\partial x_i}\right)^4 \mathrm{d}x \leqq c_3[\|f''\|_0^2 + \|u^{0''}\|_1'^2]. \tag{6.6.23}$$

Actually, the solution satisfies the elliptic system

$$\frac{\partial^2 F}{\partial\eta_i^r \partial\eta_j^s} \frac{\partial^2 u_s}{\partial x_i \partial x_j} = \frac{\partial f_i^r}{\partial x_i}, \qquad r = 1, 2, \ldots, m. \tag{6.6.24}$$

Hence it follows from (6.6.23) and (6.6.24) that

$$\int_{\mathbb{R}^3_+} \sum_{i,j,s} \left(\frac{\partial^2 u_s}{\partial x_i \partial x_j}\right)^4 \mathrm{d}x \leqq c_4[\|f\|_2^2 + \|f\|_2^4 + \|u^0\|_3'^2]. \tag{6.6.25}$$

But once we have the estimate (6.6.22), it follows from the Fourier transform technique that

$$\int_{\mathbb{R}^3_+} \sum_{r,i,j,k}' \left(\frac{\partial^3 u_r}{\partial x_i \partial x_j \partial x_k}\right)^2 \mathrm{d}x \leqq c_5[\|f\|_2^2 + \|u^0\|_3'^2], \tag{6.6.26}$$

where $\sum'$ means that u_r can be at most once differentiated with respect to x_3. Let us actually differentiate (6.6.24) with respect to x_1 or x_2. It thus follows from (6.6.25) and (6.6.26) that

(6.6.27) $$\int_{\mathbb{R}^3_+} \sum''_{r,i,j,k} \left(\frac{\partial^3 u_r}{\partial x_i \, \partial x_j \, \partial x_k} \right)^2 dx \leqq c_6[\|f\|_2^2 + \|f\|_2^4 + \|u^0\|_3'^2],$$

where $\sum''$ means the possibility of differentiating the functions u_r twice with respect to x_3. If we differentiate (6.6.24) with respect to x_3, we finally get

(6.6.28) $$\int_{\mathbb{R}^3_+} \sum_{r=1}^{m} \left(\frac{\partial^3 u_r}{\partial x_3^3} \right)^2 dx \leqq c_7[\|f\|_2^2 + \|f\|_4^2 + \|u^0\|_3'^2].$$

Actually, as far as the estimate of the first derivatives of u is concerned, this follows directly from (6.6.6); the second derivatives can be estimated from first and third derivatives by the Fourier transform technique (where we extend the solutions to $\mathbb{R}^3$).

Let us actually consider

(6.6.29) $$M = \left\{ f \in [W^{2,2}(\mathbb{R}^3_+)]^{3m} \,\middle|\, \frac{\partial f_i^r}{\partial x_i} = 0, \quad r = 1, 2, \ldots, m \right\}.$$

Let us write the system (6.6.24) in the form

(6.6.30) $$-\frac{\partial}{\partial x_i} \left[\frac{\partial F}{\partial \eta_i^r} (\nabla u) \right] = -\frac{\partial f_i^r}{\partial x_i}, \quad r = 1, 2, \ldots, m,$$

then (6.6.30) defines the operator T from $\tilde{W}$ to $(\tilde{W}/\tilde{V}) \times ([\tilde{W}^{2,2}(\mathbb{R}^3_+)]^{3m}/M) \overset{\text{def}}{=} B$ as $\left(\tilde{u}_1(.,.,0), \tilde{u}_2(.,.,0), \ldots, \tilde{u}_m(.,.,0), \frac{\partial F}{\partial \eta_1^1}(\nabla u), \ldots, \frac{\partial F}{\partial \eta_3^m}(\nabla u) \right)$. Let $K = \{\tilde{u} \in \tilde{W} \mid \|u\|_{1,\infty}' < a\}$.

Let us recall, that we speak about a diffeomorphism of an open set $K \subset \tilde{W}$ onto $T(K)$ if

(6.6.31) the mapping $\tilde{u} \mapsto T(\tilde{u})$ is one to one,

(6.6.32) $T(K)$ is open and T is a C^1 homeomorphism of K onto $T(K)$.

6.6.33. LEMMA. *The operator T defined above is a diffeomorphism of K onto $T(K)$.*

Proof. First, the operator T is one to one: if $T(\tilde{u}) = T(\tilde{v})$, then $u - v = \phi \in V + C$. We thus have

$$0 = \int_{\mathbb{R}^3_+} \left[\frac{\partial F}{\partial \eta_i^r} (\nabla u) - \frac{\partial F}{\partial \eta_i^r} (\nabla u) \right] \left(\frac{\partial u_r}{\partial x_i} - \frac{\partial v_r}{\partial x_i} \right) dx \geqq \nu \|u - v\|_1'^2,$$

hece $\phi = \text{const.}$

In a clear notation we have $T'(\tilde{u})\,\tilde{v} = \left(\tilde{v}_t, \frac{\partial^2 F}{\partial \eta_i^r \partial \eta_j^s} \frac{\partial v_s}{\partial x_j} \right)$ and we see that this differential is locally uniformly continuous in $\tilde{u}$, uniformly with respect to v on a unit ball. Let $\tilde{u}$ be fixed. We assert that $T'(\tilde{u})$ is invertible. It is sufficient to consider a solution to the problem

(6.6.34) $$v \in [W^{3,2}(\mathbb{R}^3_+)]^m, \quad v = 0 \quad \text{on } \partial\mathbb{R}^3_+,$$

(6.6.35) $$\int_{\mathbb{R}^3_+} \frac{\partial^2 F}{\partial\eta_i^r\,\partial\eta_j^s}(\nabla u)\frac{\partial v_s}{\partial x_j}\frac{\partial\phi_r}{\partial x_i}\,dx = \int_{\mathbb{R}^3_+} g_i^r \frac{\partial\phi_r}{\partial x_i}\,dx - \int_{\mathbb{R}^3_+} \frac{\partial^2 F}{\partial\eta_i^r\,\partial\eta_j^s}(\nabla u)\frac{\partial w_s}{\partial x_j}\frac{\partial\phi_r}{\partial x_i}\,dx,$$

$g \in [W^{2,2}(\mathbb{R}^3_+)]^{3m}$, $w \in [W^{3,2}(\mathbb{R}^3_+)]^m$, and to prove that

(6.6.36) $$\|v\|_3' \leqq c[\|g\|_2 + \|w\|_3'].$$

First we have

(6.6.37) $$\|v\|_1' \leqq c_1[\|g\|_0 + \|w\|_1'].$$

Put $G_i^r = g_i^r - \dfrac{\partial^2 F}{\partial\eta_i^r\,\partial\eta_j^s}(\nabla u)\dfrac{\partial w_s}{\partial x_j}$. We easily see that

(6.6.38) $$\|G\|_2 \leqq c_2[\|g\|_2 + \|w\|_3'].$$

Substitute ϕ'' for ϕ in (6.6.35) ($\phi \in [\mathscr{D}(\mathbb{R}^3_+)]^m$, of course). We then get $\left(\phi' = \dfrac{\partial\phi}{\partial x_l},\ l = 1 \text{ or } 2\right)$

(6.6.39) $$\int_{\mathbb{R}^3_+} G_i^{r''}\frac{\partial\phi_r''}{\partial x_i}\,dx = \int_{\mathbb{R}^3_+} \frac{\partial^2 F}{\partial\eta_i^r\,\partial\eta_j^s}\frac{\partial v_s''}{\partial x_j}\frac{\partial\phi_r''}{\partial x_i}\,dx + \int_{\mathbb{R}^3_+} \frac{\partial^3 F}{\partial\eta_i^r\,\partial\eta_j^s\,\partial\eta_k^t}\frac{\partial v_s'}{\partial x_j}\frac{\partial u_t'}{\partial x_k}\frac{\partial\phi_r''}{\partial x_i}\,dx.$$

We can thus substitute v for ϕ in (6.6.39). Using the extension of f to $\mathbb{R}^3$ and the Fourier transform technique, we get

(6.6.40) $$\|f\|_{L^\infty(\mathbb{R}^3_+)} \leqq \varepsilon\|f\|_{W^{2,2}(\mathbb{R}^3_+)} + \lambda(\varepsilon)\,\|f\|_{L^2(\mathbb{R}^3_+)}.$$

The inequality (6.6.15) together with (6.6.40) give

(6.6.41) $$\left(\int_{\mathbb{R}^3_+} \sum_{s,j}\left(\frac{\partial v_s'}{\partial x_j}\right)^4 dx\right)^{1/2} \leqq \varepsilon'\|v\|_3'^2 + \lambda'(\varepsilon')\,\|v\|_1'^2,$$

hence it follows from (6.6.39) and (6.6.41) that

(6.6.42) $$\|v''\|_1 \leqq \varepsilon\|v\|_3' + c_3(\varepsilon)\,\|G\|_2.$$

The equation (6.6.35) gives together with (6.6.40) for $\dfrac{\partial v_s}{\partial x_j}$ and with (6.6.41) that

(6.6.43) $$\left(\int_{\mathbb{R}^3_+} \sum_{r,i,j}\left(\frac{\partial^2 v_r}{\partial x_i\,\partial x_j}\right)^4 dx\right)^{1/2} \leqq \varepsilon\|v\|_3'^2 + c_4(\varepsilon)\,\|G\|_2^2;$$

actually, the proof is completed if we differentiate the equation $-\dfrac{\partial}{\partial x_i}\left[\dfrac{\partial^2 F}{\partial\eta_i^r\,\partial\eta_j^s}(\nabla u)\dfrac{\partial v_s}{\partial x_j}\right] = -\dfrac{\partial G_i^r}{\partial x_i}$ first with respect to x_1, x_2 and then to x_3. We thus get the inequality

(6.6.44) $$\|v\|'_3 \leqq \varepsilon \|v\|'_3 + c_5(\varepsilon) \|G\|_2 .$$

If $\varepsilon = \frac{1}{2}$, then we obtain an estimate "a priori". But this is a real estimate; this follows from the homotopy argument, if we consider the family of operators

$$(1 - t) \Delta v_r + t \frac{\partial}{\partial x_i} \left[\frac{\partial^2 F}{\partial \eta_i^r \partial \eta_j^s} (\nabla u) \frac{\partial v_s}{\partial x_j} \right], \quad 0 \leqq t \leqq 1$$

(the set of such t where (6.6.44) holds is closed and open, hence the whole [0, 1]).

6.6.45. THEOREM. *Let the operator T from $\tilde{W}$ to B, defined as above, be generated by a function F, satisfying* (6.6.6)–(6.6.8). *Let $K = \{\tilde{u} \in \tilde{W} \mid \|u\|_{1,\infty} < a\}$. Then T is a diffeomorphism of K onto $T(K)$, a homeomorphism of $\bar{K}$ onto $T(\bar{K}) = \overline{T(K)}$, a homeomorphism of ∂K onto $\partial T(K) = T(\partial K)$, and T as well as T^{-1} are bounded operators.*

Proof. First, it is clear, that $\partial K = \{\tilde{u} \in \tilde{W} \mid \|u\|'_{1,\infty} = a\}$. Using Lemma 6.6.33 for $a + \varepsilon$, where ε is small enough, we have that T is a homeomorphism of $\bar{K}$ onto $T(\bar{K})$. It is clear that T is bounded. The boundedness of T^{-1} follows from Theorem 6.6.16. From this follows that $T(\bar{K}) = \overline{T(K)}$: if $Y_k \in T(K)$ and $Y_k \to Y$ in B, then $T(\tilde{u}_k) = Y_k$ and $\tilde{u}_k$ is a bounded sequence. We can suppose that $\tilde{u}_k \rightharpoonup \tilde{u}$. But K is convex, hence weakly closed, thus $\tilde{u} \in \partial K$ if $Y \in \partial T(K)$. But, clearly, from the definition of the weak solution, $T(\tilde{u}) = Y$ follows.

Let us underline that the most important point of Theorem 6.6.45 is that T^{-1} is a bounded mapping.

6.7. A brief indroduction to the theory of isotropic polyconvex functionals in finite elasticity and the regularity of solutions for special materials

First, we recommend the reader to go back to Example 1.2.41. Here we use the notation $x \mapsto y(x)$ for the diffeomorphism of Ω onto Ω' as in 1.2.41. We refer also to the paper by GUO ZHONG-HENG [75]. For polyconvexity, we refer to the paper by J. BALL [20] and to the lecture notes by P. CIARLET [76].

We denote the tensor $\frac{\partial y_i}{\partial x_j}$ by $y\nabla$ and putting $y = x + u$ as in 1.2.41., then $y\nabla = I + u\nabla$; the Almansi tensor E defined in 1.2.41 is $E = \frac{1}{2}[u\nabla + \nabla u + \nabla u \cdot u\nabla]$ in a clear notation. We use the symbol f for the body forces introduced in 1.2.41. The symbol σ will be reserved as in 1.2.41 for the Piola stress tensor with the components σ_i^j. Writing $\sigma^j v_j = \sigma \cdot v$, we get the relation (1.2.49) in the form

(6.7.1) $$\sigma(x, \nu) = \sigma \cdot \nu .$$

The same notation brings (1.2.50) into the form

(6.7.2) $$\sigma \cdot \nabla + F = 0 .$$

Denote by T the Kirchhoff stress tensor. The relations (1.2.51) can be written in the form

(6.7.3) $$\sigma = y\nabla \cdot T ,$$

where the usual matrix product is used.

6.7.4. PROPOSITION. *Let* $y \in C^1(\Omega)$, $\det y\nabla \neq 0$. *Then there exists a unique decomposition*

(6.7.5) $$y\nabla = R \cdot U ,$$

where R is an orthogonal matrix: $R \cdot R^* = R^* \cdot R = I$, *$U$ is a symmetric positive definite matrix.*

Proof. Write $y\nabla = G$. If the decomposition (6.7.5) exists, then $G^* \cdot G = U \cdot R^* \cdot R \cdot U = U^2$; take $U = \sqrt{G^* \cdot G}$, $R = G \cdot U^{-1}$.

We call $U^2 \overset{\text{def}}{=} C$ the Cauchy-Green tensor. Since $C = \nabla y \cdot y\nabla$, we have $E = \frac{1}{2}[C - I]$. R is called the rotation tensor.

Under the corresponding regularity conditions, let us look for a path $y(t)$, $t \in [0, 1]$, of diffeomorphisms. If $\Omega \subset \mathbb{R}^3$ has a Lipschitz boundary and $\partial\Omega = \Gamma_1 \cup \Gamma_2$, then according to (1.2.55) and (1.2.56) ($\Gamma = \Gamma_2, \Gamma_1 = \partial\Omega \setminus \Gamma_2$) the work of the external forces (both $F = F(x)$ and $g = g(x)$ only, i.e., these are dead forces) is

$$\int_0^1 \mathrm{d}t \int_\Omega F \cdot \dot{y} \,\mathrm{d}x + \int_0^1 \mathrm{d}t \int_{\Gamma_2} g \cdot \dot{y} \,\mathrm{d}S = \int_\Omega Fy(1) \,\mathrm{d}x - \int_\Omega Fy(0) \,\mathrm{d}x$$
$$+ \int_{\Gamma_2} gy(1) \,\mathrm{d}S - \int_{\Gamma_2} gy(0) \,\mathrm{d}S .$$

Hyperelastic materials are supposed to possess a stored-energy function Σ (per unit volume in the reference configuration) which may be regarded as a function of any strain measure. So the increment of the stored energy is

$$\int_0^1 \mathrm{d}t \int_\Omega \dot{\Sigma} \,\mathrm{d}x = \int_\Omega (\Sigma_1 - \Sigma_0) \,\mathrm{d}x .$$

Hence the total potential energy is—apart from a constant—equal to

(6.7.6) $$\int_\Omega \Sigma \,\mathrm{d}x - \int_\Omega Fy \,\mathrm{d}x - \int_{\Gamma_2} gy \,\mathrm{d}S .$$

Actually, we shall carry out some intuitive considerations which seem to be of great importance. If Ω and the solutions y run through a set large enough, we can identify Σ in such a way that (6.7.1) as well as (6.7.2) are Euler's conditions when considering the critical points of (6.7.6). Suppose first that $\Sigma = \Sigma(E)$ (we have $\Sigma = \Sigma(E)$, $\Sigma = \Sigma(C)$, $\Sigma = \Sigma(U)$; clearly, through C the dependence is $\Sigma = \Sigma(G)$, $G = y\nabla$). Then $\sigma \cdot \nu = G \cdot \frac{d\Sigma}{dE} \cdot \nu$ on Γ_2 $\left(\frac{d\Sigma}{dE}\right.$ is the tensor $\left.\frac{\partial \Sigma}{\partial E_{ij}}\right)$. So, since Γ_2 can change, we get $\sigma = G \cdot \frac{d\Sigma}{dE}$, hence $G^{-1} \cdot \sigma = \frac{d\Sigma}{dE}$. But it follows from (6.7.3) that $G^{-1}\sigma = T$. "hence" $T = \frac{d\Sigma}{dE}$. We have thus come to the Hook's law (1.2.53). The relation (6.7.2) follows automatically. It is natural to suppose that the relation

(6.7.7) $$\frac{d\Sigma}{dE} = T$$

for $E = \frac{1}{2}[C - I]$, where C are symmetric positive definite tensors, is one-to-one. It is also quite natural to suppose that $\Sigma(E)$ is positive definite on the set mentioned. The most important reason for this is the relation that we have obtained,

(6.7.8) $$\delta \int_\Omega \Sigma \, dx = \int_\Omega \operatorname{tr}\left(\frac{d\Sigma}{dE} \cdot \delta E\right) dx = \int_\Omega \operatorname{tr}(T \cdot \delta E) \, dx.$$

For the same reason we can write

(6.7.9) $$\delta \int_\Omega \Sigma \, dx = \int_\Omega \operatorname{tr}\left(\frac{d\Sigma}{dU} \cdot \delta U\right) dx$$

and putting $S = \frac{d\Sigma}{dU}$, we get by simple calculations that

(6.7.10) $$S = \tfrac{1}{2}(Tu + UT).$$

As far as symmetric positive definite tensors are concerned, it is thus reasonable to suppose that the relation

(6.7.11) $$\frac{d\Sigma}{dU} = S$$

is one-to-one or, more over, that $\Sigma(U)$ is positive definite on the set mentioned.

The symmetric tensor S is called Jaumann's tensor.

Let us remark that, since $\Sigma = \Sigma(U)$, it must be for an arbitrary orthogonal matrix O (we write $\Sigma = \Sigma(F)$)

(6.7.12) $$\Sigma(F) = \Sigma(OF),$$

which is called the objectivity of Σ. Clearly, if (6.7.12) is true, then $\Sigma = \Sigma(U)$.

The stored-energy function is called isotropic if for all orthogonal matrices O

(6.7.13) $$\Sigma(F) = \Sigma(OFO^*).$$

If Σ is written as $\Sigma = \Sigma(C)$, then the relation (6.7.13) implies

(6.7.14) $$\Sigma(C) = \Sigma(OCO^*) = \Sigma(\lambda_1, \lambda_2, \lambda_3),$$

where λ_i are the eigenvalues of the matrix C. For the proof, see, for example, C. TRUESDELL [77]. Let us now consider only the isotropic case. We have

$\lambda_i = \lambda_i(I_1, I_2, I_3)$ where I_j are the invariants of the matrix C defined, as usual, as

(6.7.15) $$\det(C - \lambda I) = -\lambda^3 + \lambda^2 I_1 - \lambda I_2 + I_3,$$
$$I_1 = \lambda_1 + \lambda_2 + \lambda_3 = \operatorname{tr} C,$$
$$I_2 = \lambda_1\lambda_2 + \lambda_1\lambda_3 + \lambda_2\lambda_3 = \operatorname{tr} \operatorname{adj} C,$$

where for a nonsingular matrix

(6.7.16) $$\operatorname{adj} A \overset{\text{def}}{=} \det A \cdot (A)^{-1}.$$

$I_3 = \det C$. We have

(6.7.17) $$\operatorname{adj} A^* = (\operatorname{adj} A)^*, \qquad \operatorname{adj} B \cdot C = \operatorname{adj} C \cdot \operatorname{adj} B.$$

In the general case, we define adj A as the matrix of the cofactors of the transposed matrix A^*.

But $\operatorname{tr} C = \operatorname{tr} G^*G = \frac{\partial y_i}{\partial x_j}\frac{\partial y_i}{\partial x_j} \overset{\text{def}}{=} |G|^2$, $\det C = (\det G)^2$, $\operatorname{tr}\operatorname{adj} C = \operatorname{tr}\operatorname{adj} G^*G = \operatorname{tr}\operatorname{adj} G^* \cdot \operatorname{adj} G = |\operatorname{adj} G|^2$, hence in the isotropic case,

(6.7.18) $$\Sigma = \Sigma(|G|, |\operatorname{adj} G|, \det G)$$

6.7.19. EXAMPLE. The Mooney-Rivlin material.

$$\Sigma = c_1(I_1 - 3) + c_2(I_2 - 3) = c_1(|G|^2 - 3) + c_2(|\operatorname{adj} G|^2 - 3),$$
$$c_1 > 0,\ c_2 > 0.$$

The Odgen material:

$$\Sigma = c_1|G|^2 + c_2|\operatorname{adj} G|^2 + \Gamma(\det G), \quad c_1 > 0,\ c_2 > 0,$$

Γ is a twice continuously differentiable function in $\mathbb{R}^1_+$ such that $\Gamma''(s) \geqq 0$, $\lim_{s \to +0} \Gamma(s) = \infty$.

The most difficult problem of finite elasticity is to find reasonable conditions for Σ which guarantee the existence of weak or strong solutions to the problem (1.2.54), (1.2.55), (1.2.56). If we write $\Sigma = \Sigma(y\nabla) = \Sigma(\eta)$, then the condition

$$\frac{\partial^2 \Sigma}{\partial \eta_i^r \partial \eta_j^s}(\eta)\, \zeta_i^r \zeta_j^s \geqq c|\zeta|^2 \quad \text{is not realistic.}$$

One of the most promising methods is described in the paper [20]. This method seems to be very good for isotropic materials, which we consider in our lecture notes. We refer the reader to the work [20]; here we shall be interested only in some basic notions and results from [20] and we add some small improvements. In (6.7.8) let us write $\eta \in \mathbb{R}^9$ instead of $y\nabla$, $\zeta \in \mathbb{R}^9$ instead of adj $y\nabla$, $\delta \in \mathbb{R}^1_+$ (sometimes in $\mathbb{R}^1$) instead of det $y\nabla$, and $\Sigma = \Sigma(|\eta|, |\zeta|, \delta)$. In general, we can consider $\Sigma = \Sigma(\eta, \zeta, \delta)$. To give precision to our notation, in the next we shall write $\Sigma = \Sigma(\eta)$ only and we shall assume the existence of a function $G(\eta, \zeta, \delta)$, continuous on $\mathbb{R}^9 \times \mathbb{R}^9 \times \mathbb{R}^1_+$ and such that

(6.7.20)′ $$\Sigma(\eta) = G(\eta, \operatorname{adj} \eta, \det \eta).$$

We assume the following growth conditions:

(6.7.20) $$|G(\eta, \zeta, \delta)| \leqq c(1 + |\eta|^p + |\zeta|^q + \delta^r + \delta^{-s})$$

with $p \geqq 2$, $q \geqq p'$, $r > 1$, $s > 0$, and coerciveness conditions

(6.7.21) $$G(\eta, \zeta, \delta) \geqq c_1(|\eta|^p + |\zeta|^q + \delta^r + \delta^{-s}) - c_2, \quad c_1 > 0$$

if G depends on δ or

(6.7.22) $$G(\eta, \zeta) \geqq c_1(|\eta|^p + |\zeta|^q) - c_2$$

if G does not depend on δ.

Let, as usual, $\Omega \subset \mathbb{R}^3$ be a bounded domain with a Lipschitz continuous boundary, let $F \in [L^{p'}(\Omega)]^3$, $g \in [L^{p'}(\Gamma_2)]^3$, where $\partial\Omega = \Gamma_1 \cup \Gamma_2 \cup M$ with $|M|_{n-1} = 0$, Γ_i be open sets (possibly empty). Let $\psi^0 \in [W^{1,p}(\Omega)]^3$. Define

(6.7.23) $$\Phi(\psi) = \int_\Omega G(\psi\nabla, \operatorname{adj}\psi\nabla, \det\psi\nabla)\, dx - \int_\Omega F \cdot \psi\, dx - \int_{\Gamma_2} g \cdot \psi\, dS$$

on the set

(6.7.24) $$U = \{\psi \in [W^{1,p}(\Omega)]^3 \mid \operatorname{adj}\psi\nabla \in [L^q(\Omega)]^9,\ \det\psi\nabla \in L^r(\Omega),$$
$$\psi = \psi^0 \text{ on } \Gamma_1,\ \det\psi\nabla > 0 \text{ almost everywhere in } \Omega \text{ in the case of (6.7.21)}\}$$

provided that $\psi^0 \in U$ and $\Phi(\psi^0) < \infty$.

6.7.25. DEFINITION. *The weak solution of finite elasticity is a function $\psi \in U$ minimizing Φ over U.*

6.7.26. DEFINITION. *A function $\Sigma(\eta)$ is called polyconvex, if there exists a continuous function $G(\eta, \zeta, \delta)$ in $\mathbb{R}^9 \times \mathbb{R}^9 \times \mathbb{R}^1_+$, satisfying* (6.7.20)′ *and convex here, i.e. in the space $\mathbb{R}^{19}_+$.*

6.7.27. LEMMA. *Let $\psi^t \rightharpoonup \psi$ in $[W^{1,p}(\Omega)]^3$ and let* $\operatorname{adj}\psi^t\nabla \rightharpoonup H$ *in $[L^q(\Omega^9)]$. Then $H = \operatorname{adj}\psi\nabla$.*

Proof. First we have the important formula

(6.7.28) $$(\operatorname{adj}\psi\nabla)_{ij} = \frac{1}{2}\varepsilon_{ikl}\varepsilon_{jmn}\frac{\partial\psi_m}{\partial x_k}\frac{\partial\psi_n}{\partial x_l};$$

ε_{ikl} is the Levi-Civita tensor.

Let $\psi \in [C^\infty(\bar{\Omega})]^3$, $\phi \in \mathscr{D}(\Omega)$. Then

(6.7.29) $$\int_\Omega (\operatorname{adj}\psi\nabla)_{ij}\phi\, dx = -\frac{1}{2}\int_\Omega \varepsilon_{ikl}\varepsilon_{jmn}\psi_m \frac{\partial^2\psi_n}{\partial x_k\, \partial x_l}\phi\, dx - \frac{1}{2}\int_\Omega \varepsilon_{ikl}\varepsilon_{jmn}\psi_m \frac{\partial\psi_n}{\partial x_l}\frac{\partial\phi}{\partial x_k}\, dx.$$

But the first integral on the right-hand side of (6.7.29) is zero. Now, for $\psi \in [W^{1,p}(\Omega)]^3$ there exists a sequence $\psi^k \to \psi$, $\psi^k \in [C^\infty(\bar{\Omega})]^3$. Thus

(6.7.30) $$\int_\Omega (\operatorname{adj}\psi\nabla)_{ij}\phi\, dx = -\frac{1}{2}\int_\Omega \varepsilon_{ikl}\varepsilon_{jmn}\psi_m \frac{\partial\psi_n}{\partial x_l}\frac{\partial\phi}{\partial x_k}\, dx$$

$\forall\psi \in [W^{1,p}(\Omega)]^3$ $(p \geqq 2!)$.

If ψ^t is the sequence in question, we have because $W^{1,p}(\Omega) \subsetneq L^p(\Omega)$ is a completely continuous mapping, that

(6.7.31)
$$-\frac{1}{2}\int_\Omega \varepsilon_{ikl}\varepsilon_{jmn}\psi_m^t \frac{\partial \psi_n^t}{\partial x_l}\frac{\partial \phi}{\partial x_k}\,dx$$
$$\to -\frac{1}{2}\int_\Omega \varepsilon_{ikl}\varepsilon_{jmn}\psi_m \frac{\partial \psi_n}{\partial x_k}\frac{\partial \phi}{\partial x_k}\,dx,$$

on the other hand,

(6.7.32)
$$\lim_{t\to\infty}\int_\Omega (\operatorname{adj}\psi^t\nabla)_{ij}\phi\,dx = \int_\Omega H_{ij}\phi\,dx$$

but by virtue of (6.7.30), (6.7.31), and (6.7.32) we have $(\operatorname{adj}\psi\nabla)_{ij} = H_{ij}$.

Let us mention the trivial fact, that for $\psi \in [W^{1,p}(\Omega)]^3$, $\det\psi\nabla$ is defined by the usual definition almost everywhere and it is a measurable function, in general not in $L^1(\Omega)$ or even not a distribution.

On the other hand

6.7.33. LEMMA. *Let* $\psi \in [W^{1,p}(\Omega)]^3$, $\operatorname{adj}\psi\nabla \in [L^q(\Omega)]^9$, $p \geqq 2$, $q \geqq p'$. *Then* $\det\psi\nabla \in L^1(\Omega)$ *and* $\forall\phi \in \mathscr{D}(\Omega)$

(6.7.34)
$$\int_\Omega \det\psi\nabla\cdot\phi\,dx = -\int_\Omega \psi_1(\operatorname{adj}\psi\nabla)_{i1}\frac{\partial\phi}{\partial x_i}\,dx.$$

Proof. Put $(\operatorname{adj}\psi\nabla)_{i1} = w_i$, take $\mu \in \mathscr{D}(\Omega)$. We have

(6.7.35)
$$\int_\Omega w_i\frac{\partial\mu}{\partial x_i}\,dx = \frac{1}{2}\int_\Omega \varepsilon_{ikl}\varepsilon_{1mn}\frac{\partial\psi_m}{\partial x_k}\frac{\partial\psi_n}{\partial x_l}\frac{\partial\mu}{\partial x_i}\,dx.$$

Let $\psi^k \to \psi$ in $[W^{1,p}(\Omega)]^3$, $\psi^k \in [C^\infty(\bar\Omega)]^3$ which can be done in virtue of Theorem 2.1.9.

Putting ψ^k instead of ψ in (6.7.35), we get $-\int_\Omega \frac{\partial w_i^k}{\partial x_i}\mu\,dx = 0$. Thus

(6.7.36)
$$\int_\Omega w_i\frac{\partial\mu}{\partial x_i}\,dx = 0.$$

Consider

$$\int_\Omega \psi_1^k(\operatorname{adj}\psi\nabla)_{i1}\frac{\partial\phi}{\partial x_i}\,dx = \int_\Omega(\operatorname{adj}\psi\nabla)_{i1}\frac{\partial(\phi\psi_1^k)}{\partial x_i}\,dx - \int_\Omega(\operatorname{adj}\psi\nabla)_{i1}\phi\frac{\partial\psi_1^k}{\partial x_i}\,dx$$
$$= -\int_\Omega(\operatorname{adj}\psi\nabla)_{i1}\phi\frac{\partial\psi_1^k}{\partial x_i}\,dx.$$

Then

(6.7.37)
$$\int_\Omega \psi_1^k(\operatorname{adj}\psi\nabla)_{i1}\frac{\partial\phi}{\partial x_i}\,dx = -\int_\Omega(\operatorname{adj}\psi\nabla)_{i1}\phi\frac{\partial\psi_1^k}{\partial x_i}\,dx.$$

But we can pass to the limit in (6.7.37) as $k\to\infty$ and we get the result.

6.7.38. LEMMA. *Let* $\psi^k \rightharpoonup \psi$ *in* $[W^{1,p}(\Omega)]^3$, $\operatorname{adj} \psi^k\nabla \rightharpoonup H$ *in* $[L^q(\Omega)]^9$, *and* $\det \psi^k\nabla \rightharpoonup h$ *in* $L^r(\Omega)$. *Then* $h = \det \psi\nabla$.

Proof. According to (6.7.34) we have

$$\int_\Omega \det \psi^k\nabla \cdot \phi \,\mathrm{d}x = -\int_\Omega \psi_1^k(\operatorname{adj} \psi^k\nabla)_{i1} \frac{\partial \phi}{\partial x_i} \,\mathrm{d}x.$$

But $\psi_1^k \to \psi_1$ in $L^p(\Omega)$ and we thus get by virtue of Lemmas 6.7.27 and 6.7.33 that

$$\int_\Omega h\phi \,\mathrm{d}x = -\int_\Omega \psi_1(\operatorname{adj} \psi\nabla)_{i1} \frac{\partial \phi}{\partial x_i} \,\mathrm{d}x = \int_\Omega \det \psi\nabla \cdot \phi \,\mathrm{d}x.$$

6.7.39. THEOREM. *Suppose that the stored energy function* Σ *is polyconvex and satisfies* (6.7.20), (6.7.21), *or* (6.7.22). *Then there exists a weak solution* ψ *provided that* $\Gamma_1 \neq \emptyset$. *If* (6.7.21) *is valid then* $\det \psi\nabla > 0$ *almost everywhere in* Ω.

Proof. We shall consider the more complicated case (6.7.21) only. First, it follows from (6.7.21) that there exists $R > 0$ such that if $\psi \in U$ and $\|\psi\|_{[W^{1,p}(\Omega)]^3} + \|\operatorname{adj} \psi\nabla\|_{[L^q(\Omega)]^9} + \|\det \psi\nabla\|_{L^r(\Omega)} \geqq R$, then $\Phi(\psi) > \Phi(\psi_0)$. Define $\tilde{G}(\eta, \zeta, \delta) = G(\eta, \zeta, \delta)$ for $\delta > 0$ and $\tilde{G}(\eta, \zeta, 0) = \infty$. Let $a = \inf_{\varphi \in U} \Phi(\psi)$, where U is the set (6.7.24), and let $\psi^k \in U$ be such that $\Phi(\psi^k) \searrow a$. We have

$$(6.7.40) \qquad \|\psi^k\|_{[W^{1,p}(\Omega)]^3} + \|\operatorname{adj} \psi^k\nabla\|_{[L^q(\Omega)]^9} + \|\det \psi^k\nabla\|_{L^r(\Omega)} < R.$$

We can suppose $\psi^k \rightharpoonup \psi$ in $[W^{1,p}(\Omega)]^3$, $\operatorname{adj} \psi^k\nabla \rightharpoonup H$ in $[L^q(\Omega)]^9$, $\det \psi\nabla \rightharpoonup h$ in $L^r(\Omega)$.

Actually, we use Mazur's lemma for the triple $(\psi^k, \operatorname{adj} \psi^k\nabla, \det\psi^k\nabla) \overset{\text{def}}{=} f^k$. We thus get convex combinations

$$m^k = \sum_{i=k}^{n_k} \lambda_i f^i, \qquad \sum_{i=k}^{n_k} \lambda_i = 1, \quad 0 \leqq \lambda_i,$$

such that $m^k \to (\psi, \operatorname{adj} \psi\nabla, \det \psi\nabla)$ in $[W^{1,p}(\Omega)]^3 \times [L^q(\Omega)]^9 \times L^r(\Omega)$ and almost everywhere in Ω (by virtue of Lemmas 6.7.27 and 6.7.38). The convexity of $G(\eta, \zeta, \delta)$ implies

$$(6.7.41) \qquad \int_\Omega G(m_1^k, m_2^k, m_3^k) \,\mathrm{d}x \leqq \int_\Omega G(\psi^k, \operatorname{adj} \psi^k\nabla, \det \psi^k\nabla) \,\mathrm{d}x.$$

In virtue of (6.7.21) we can use Fatou's lemma and we get

$$(6.7.42) \qquad \int_\Omega \bar{G}(\psi, \operatorname{adj} \psi\nabla, \det \psi\nabla) \,\mathrm{d}x \leqq a.$$

The inequality (6.7.42) first implies that $\det \psi\nabla(x) > 0$ almost everywhere in Ω. Hence $\psi \in U$ and by virtue of (6.7.42), ψ is a minimizing point.

6.7.43. THEOREM. *Suppose that all the conditions from Theorem* 6.7.39 *but that concerned with* Γ_1 *are satisfied. Let* $\Gamma_1 = \emptyset$. *Then there exists a weak solution* ψ *iff*

$$(6.7.44) \qquad \int_\Omega F \,\mathrm{d}x + \int_{\partial\Omega} g \,\mathrm{d}S = 0.$$

Every function ψ + const *is also a solution. If* (6.7.21) *is valid then* $\det \psi\nabla > 0$ *almost everywhere in* Ω.

Proof. Clearly, if (6.7.44) is not satisfied, then $\inf \Phi(\psi^0 + \text{const}) = -\infty$. If (6.7.44) is true then, considering the space $[W^{1,p}(\Omega)]^3/C$, we get the result in the same way as in the previous theorem.

6.7.45. REMARK. If ψ is a solution under the conditions of Theorem 6.7.43, then

$$\int_{\Omega} F \times \psi \, dx + \int_{\partial\Omega} g \times \psi \, dS = 0, \tag{6.7.46}$$

where $F \times \psi$ and $g \times \psi$ mean vector products (equilibrium of momenta). One can get (6.7.46) using the objectivity condition (6.7.12).

6.7.47. THEOREM. *Let* $\Sigma = \Sigma(I_1^{\frac{1}{1}}, I_2^{\frac{1}{2}}, I_3^{\frac{1}{3}})$. *Let* $\Sigma(\xi_1, \xi_2, \xi_3)$ *be a continuous function in* $\overline{\mathbb{R}}_+^1 \times \overline{\mathbb{R}}_+^1 \times \mathbb{R}_+^1$, *non-decreasing in* ξ_1 *and in* ξ_2 *and convex in* (ξ_1, ξ_2, ξ_3). *Then* Σ *is polyconvex.*

Proof. Put $G(\eta, \zeta, \delta) = \Sigma(|\eta|, |\zeta|, \delta)$. Then

$$\begin{aligned} &\Sigma(|\alpha\eta^1 + (1-\alpha)\eta^2|, |\alpha\zeta^1 + (1-\alpha)\zeta^2|, \alpha\delta^1 + (1-\alpha)\delta^2) \\ &\leqq \Sigma(\alpha|\eta^1| + (1-\alpha)|\eta^2|, \alpha|\zeta^1| + (1-\alpha)|\zeta^2|, \alpha\delta^1 + (1-\alpha)\delta^2) \\ &\leqq \alpha\Sigma(|\eta^1|, |\zeta^1|, \delta^1) + (1-\alpha)\Sigma(|\eta^2|, |\zeta^2|, \delta^2) \end{aligned}$$

with $0 \leqq \alpha \leqq 1$.

6.7.48. EXAMPLE. Consider the stored energy function in the form of the Saint-Venant-Kirchhoff law:

$$I = \frac{\lambda}{2}(\operatorname{tr} E)^2 + \mu \operatorname{tr} E^2, \quad \lambda \geqq 0, \ \mu > 0$$

are constants. We get

$$\Sigma = \frac{\lambda + 2\mu}{8}|G|^4 - \frac{3\lambda + 2\mu}{4}|G|^2 - \frac{\mu}{2}|\operatorname{adj} G|^2 + \frac{3}{4}\mu + \frac{9}{8}\lambda. \tag{6.7.49}$$

Here we used the fact that $\operatorname{tr} C^2 = \lambda_1^2 + \lambda_2^2 + \lambda_3^2 = I_1^2 - 2I_2$. In the representation (6.7.49), this stored energy function is not polyconvex. We can also see easily that the mapping $\dfrac{d\Sigma}{dE} = T$ is one-to-one but the mapping $\dfrac{d\Sigma}{dU} = S$ is not.

6.7.50. EXAMPLE. Take 6.7.19 into account. Clearly Σ are polyconvex. For the Mooney-Rivlin material (the author has not considered the Ogden material), $\Sigma = \Sigma(E)$ is positive definite and $\Sigma = \Sigma(U)$ is also positive definite.

6.7.51. EXAMPLE. Let us take a special case of the Ogden material into account:

(6.7.52)
$$a|G|^2 + b|\text{adj } G|^2 + c(\det G)^2 + d(\det G)^{-2k},$$
$$k > 0,\ a > 0,\ b > 0,\ c > 0,\ d > 0.$$

Considering the Saint-Venant-Kirchhoff stored energy function Σ, we get for the Taylor expansion in ∇u (for ∇u we turn back to the classical notation) that $\frac{\lambda}{2}(\text{tr } E)^2 + \mu \text{ tr } E^2 \doteq \frac{\lambda}{2}(\text{div } u)^2 + \mu e_{ij}e_{ij}$, where $e_{ij} = \frac{1}{2}\left(\frac{\partial u_i}{\partial x_j} + \frac{\partial u_j}{\partial x_i}\right)$. Let us calculate such a Taylor expansion for (6.7.52). We get in such a way that if $\lambda > 0$, if

(6.7.53)
$$\frac{\mu}{2} - \frac{\lambda}{4} < a < \frac{\mu}{2}, \qquad b = \frac{\mu}{2} - a,$$
$$c = \frac{\lambda - 4\mu k - 2\mu + 4a + 4ak}{4(1 + k)}, \qquad d = \frac{\lambda + 2\mu}{4k(1 + k)},$$

and if k is small enough, then

$$a|G|^2 + b|\text{adj } G|^2 + c(\det G)^2 + d(\det G)^{-2k} \doteq \frac{\lambda}{2}(\text{div } u)^2 + \mu e_{ij}e_{ij}.$$

This means that, asymptotically in ∇u near the origin, the Saint-Venant-Kirchhoff stored energy is equivalent to the Ogden stored energy (of a special form).

6.7.54. REMARK. As discussed in detail in the work [20], we can look for a minimum of the functional (6.7.23) provided that $\det \psi\nabla = 1$ which corresponds to the incompressibility. We get the existence easily as in the second case ((6.7.21)) of Theorem 6.7.39. In virtue of Lemma 6.7.38, the constraint $\det \psi\nabla = 1$ is weakly closed.

6.7.55. REMARK. The reader can readily study the easier case when $G(\eta, \zeta, \delta)$ is defined on $\mathbb{R}^9 \times \mathbb{R}^9 \times \mathbb{R}^1$ and is convex here with appropriate growth conditions: no condition for $\delta \to 0$ is imposed.

Actually, we shall apply the results of Section 6.6. to special Ogden materials as described in Example 6.7.51. We shall, however, do the following deceit: because the regularity estimates are obtained for y near identity "thus" we can study, instead of (6.7.52), the Saint-Venant-Kirchoff stored energy. Clearly, this is a weak point of our considerations. In the sequel we use the displacement vector u and the standard notation as ∇u etc.

Having thus, as mentioned in 1.2.41 and in this section, the components of E:

(6.7.56)
$$2\varepsilon_{ij} = \frac{\partial u_i}{\partial x_j} + \frac{\partial u_j}{\partial x_i} + \frac{\partial u_k}{\partial x_i}\frac{\partial u_k}{\partial x_j},$$

we shall consider the stored energy function

(6.7.57)
$$\Sigma(E) = \frac{\lambda}{2}\varepsilon_{kk}^2 + \mu\varepsilon_{ij}\varepsilon_{ij}.$$

Let us put, as usual,

$$\mu = \frac{E}{2(1+\sigma)}, \qquad \lambda = \frac{E\sigma}{(1+\sigma)(1-2\sigma)}, \tag{6.7.58}$$

where E is Young's modulus and σ is Poisson's ratio, $0 \leqq \sigma < \frac{1}{2}$, $E > 0$. Put $\varepsilon_{ij} = e_{ij} + A_{ij}$ and write $\delta e_{ij} \overset{\text{def}}{=} \frac{\mathrm{d}}{\mathrm{d}t}(e_{ij}(u + t\phi))\,|_{t=0} = \tilde{e}_{ij}$ etc. Considering Euler's equation for the functional

$$\Phi(u) = \int_{\mathbb{R}^3_+} [\Sigma(E) - F_i u_i]\,\mathrm{d}x \tag{6.7.59}$$

and the displacement problem with u prescribed on ∂R^3_+, we can study another stored energy function which leads to the same Euler equations:

$$\Sigma(\nabla u) = \mu\left[\frac{1}{2}|\nabla u|^2 + \frac{1}{2}(\operatorname{div} u)^2 + 2e_{ij}A_{ij} + A_{ij}A_{ij}\right] + \frac{\lambda}{2}\left[\operatorname{div} u + \frac{1}{2}|\nabla u|^2\right]^2. \tag{6.7.60}$$

We shall work with (6.7.60) in the sequel.

We have

$$\delta^2\Sigma = \lambda[\operatorname{div} u + \tfrac{1}{2}|\nabla u|^2]\,|\nabla\phi|^2 + \lambda[\operatorname{div}\phi + (\nabla u, \nabla\phi)]^2 + \mu[|\nabla\phi|^2 + (\operatorname{div}\phi)^2 + 4\tilde{e}_{ij}\tilde{A}_{ij} + 2e_{ij}\tilde{\tilde{A}}_{ij} + 2\tilde{A}_{ij}\tilde{A}_{ij} + A_{ij}\tilde{\tilde{A}}_{ij}]. \tag{6.7.61}$$

We further have

$$\tilde{A}_{ij} = \frac{1}{2}\left(\frac{\partial\phi_k}{\partial x_i}\frac{\partial u_k}{\partial x_j} + \frac{\partial u_k}{\partial x_i}\frac{\partial\phi_k}{\partial x_j}\right), \tag{6.7.62}$$

$$\tilde{\tilde{A}}_{ij} = \frac{\partial\phi_k}{\partial x_i}\frac{\partial\phi_k}{\partial x_j}. \tag{6.7.63}$$

It is $A_{ij}\tilde{\tilde{A}}_{ij} \geqq 0$, $|\tilde{e}_{ij}\tilde{A}_{ij}| \leqq |\nabla\phi|^2\,|\nabla u|$, $|e_{ij}\tilde{\tilde{A}}_{ij}| \leqq |\nabla u|\,|\nabla\phi|^2$, hence we get from (6.7.61) that

$$\delta^2 F \geqq \lambda[\tfrac{1}{2}|\nabla u|^2 - |\operatorname{div} u|]\,|\nabla\phi|^2 + \mu|\nabla\phi|^2[1 - 6|\nabla u|]. \tag{6.7.64}$$

If $\frac{\partial u_r}{\partial x_i} = \eta_i^r$ we thus get

$$\frac{\partial^2\Sigma}{\partial\eta_i^r\,\partial\eta_j^s}(\nabla u)\,\zeta_i^r\zeta_j^s \geqq \frac{E}{2(1+\sigma)(1-2\sigma)}[1 - 2\sigma - 6a + 12a\sigma - 2\sqrt{3}\sigma a]\,|\zeta|^2. \tag{6.7.65}$$

We further get from (6.7.60):

$$\delta^4\Sigma = (3\lambda + 6\mu)\,|\nabla\phi|^4. \tag{6.7.66}$$

We therefore obtain

6.7.67. LEMMA. *On the set* $|\nabla u| < a$, *the coefficient* ν *of the ellipticity* (6.6.6) *is estimated by the constant from* (6.7.65), *hence* $\nu > 0$ *if*

(6.7.68) $$a < \frac{1 - 2\sigma}{6 - \sigma(12 - 2\sqrt{3})}.$$

The condition (6.6.8) *is satisfied if*

(6.7.69)

$$a < \frac{-3 + 6\sigma - \sigma\sqrt{3} + \sqrt{(-3 + 6\sigma - \sigma\sqrt{3})^2 + 18(1 - 2\sigma)(1 - \sigma)}}{18(1 - \sigma)}.$$

For $\sigma = 0$, the condition (6.7.68) is satisfied if $a < \frac{1}{b} \doteq 0.167$, the condition (6.7.69) if $a < 0.122$, for $\sigma = \frac{1}{3}$ if $a < 0.106$ and $a < 0.081$, respectively.

6.7.70. THEOREM. *If a satisfies* (6.7.69) *then Theorem* 6.6.45 *can be applied to F substituted by* (6.7.60).

For more detailed considerations concerned with the finite elasticity with (6.7.60), we refer the reader to [70].

The significance of Theorem 6.7.70 shows off if the data paths as well as the solution paths are considered: In our notation, let $X(t) \in C([0, 1], B)$ (the data path). Suppose that $X(0) = 0$ and let $\tilde{u}(0) = 0$ be the corresponding solution. We look for $\tilde{u}(t) \in C([0, t_0], \tilde{W})$ such that $T(\tilde{u}(t)) = X(t)$. It follows from Theorem 6.7.70 (with small variations of Theorem 6.6.33) that such a solution uniquely exists in $\tilde{K} = \{\tilde{u} \in \tilde{W} \mid \|\nabla u\|'_{1,\infty} < a\}$ and that $u(t) \in K$ and $u(t_0) \in \partial K$ for $0 \leqq t < t_0 \leqq 1$. If, in fact, on the set $|\eta| \leqq a$ the response function $\Sigma(\eta)$ is defined, then $u(t_0) \in \partial K$ means that the solution path did reach the boundary of the domain of definition of the stored energy function

6.8. Nonlinear elasticity

We shall modify slightly Theorem 6.6.45 for its application to elasto-plasticity. We recommend the reader to recall the introduction to nonlinear elasticity from 1.2. before reading this section.

So we consider the small strain tensor

(6.8.1) $$e_{ij} = \frac{1}{2}\left(\frac{\partial u_i}{\partial x_j} + \frac{\partial u_j}{\partial x_i}\right),$$

which is the linear part of the Almansi tensor (1.2.43). We denote by σ_{ij} the Cauchy stress tensor and by F the body force. The equilibrium condition has the form (1.2.64). Let us denote by

(6.8.2) $$\sigma^*_{ij} = \sigma_{ij} - \tfrac{1}{3}\delta_{ij}\sigma_{kk}, \qquad e^*_{ij} = e_{ij} - \tfrac{1}{3}\delta_{ij}e_{kk}$$

the deviators of the tensors considered and write

(6.8.3) $$S \overset{\text{def}}{=} \sigma^*_{ij}\sigma^*_{ij}, \qquad \Gamma \overset{\text{def}}{=} e^*_{ij}e^*_{ij}.$$

Let us consider a function $\mu(\Gamma)$ once continuously differentiable in $\overline{\mathbb{R}_1^+}$. The stress-strain relation of Hencky-Mises can be written as

(6.8.4) $$\sigma_{ij} = [k - \tfrac{2}{3}\mu(\Gamma)]\,\delta_{ij}e_{kk} + 2\mu(\Gamma)\,e_{ij},$$

where k is the bulk modulus and $\mu(\Gamma)$ is the Lamé function. We suppose that

(6.8.5) $$0 < \mu_0 \leqq \mu(\Gamma) \leqq \tfrac{3}{2}k.$$

We get from (6.8.4) that

(6.8.6) $$\sigma_{kk} = 3ke_{kk},$$

(6.8.7) $$\sigma_{ij}^* = 2\mu e_{ij}^*, \qquad S = 4\mu^2\Gamma.$$

Put $s = \sqrt{S}$, $\gamma = \sqrt{\Gamma}$. Then

(6.8.8) $$s = 2\mu\gamma.$$

A physically natural condition is that the function $s = s(\gamma)$ is increasing. We have $s' = 2[2\mu'\Gamma + \mu]$, so we shall suppose that

(6.8.9) $$0 < \mu_1 \leqq \mu + 2\mu'\Gamma \leqq \mu_2 < \infty.$$

Let $\Omega \subset \mathbb{R}^3$ be a domain with a Lipschitz boundary, $\partial\Omega = \Gamma_\sigma \cup \Gamma_u$ ($\cup$ a set of zero surface measure). Suppose that $\Gamma_u \neq \emptyset$, $F \in [L^2(\Omega)]^3$, $g \in [L^2(\Gamma_\sigma)]^3$, $u^0 \in [W^{1,2}(\Omega)]^2$.

A weak solution to elasto-plasticity is $u \in [W^{1,2}(\Omega)]^3$ such that

(6.8.10) $$u = u^0 \quad \text{on } \Gamma_u,$$

$\forall v \in [W^{1,2}(\Omega)]^3$ such that $v = 0$ on Γ_u:

(6.8.11) $$\int_\Omega [(k - \tfrac{2}{3}\mu(\Gamma(u)))\,e_{kk}(u)\,e_{kk}(v) + 2\mu(\Gamma(u))\,e_{ij}(u)\,e_{ij}(v)]\,dx$$
$$= \int_\Omega F_i v_i\,dx + \int_{\Gamma_\sigma} g_i v_i\,dS.$$

Put $\Gamma(u, v) = e_{ij}^*(u)\,e_{ij}^*(v)$. Then we can write the left-hand side of (6.8.11) in the form

(6.8.12) $$\int_\Omega [ke_{ll}(u)\,e_{mm}(v) + 2\mu(\Gamma(u))\,\Gamma(u, v)]\,dx;$$

of course, $\Gamma(u) = \Gamma(u, u)$.

Let

(6.8.13) $$\Phi(u) \overset{\text{def}}{=} \int_\Omega [\tfrac{1}{2}k|e_{ll}(u)|^2 + \tfrac{1}{2}\int_0^{\Gamma(u)} \mu(t)\,dt - F_i u_i]\,dx - \int_{\Gamma_\sigma} g_i u_i\,dS.$$

6.8.14. THEOREM. *For $v \in [W^{1,2}(\Omega)]^3$, $v = 0$ on Γ_u, we have*

(6.8.15) $$m\|v\|^2_{[W^{1,2}(\Omega)]^3} \leqq D^2\Phi(u, v, v) \leqq M\|v\|^2_{[W^{1,2}(\Omega)]^3}.$$

For the proof, see [19]; for v in question we use Korn's inequality

$$\int_\Omega e_{ij}(v)\,e_{ij}(v)\,dx \geqq c_1\|v\|^2_{[W^{1,2}(\Omega)]^3}.$$

6.8.16. THEOREM. *There exists a unique weak solution of elasto-plasticity and the mapping $[F, u^0] \to u$ is a Lipschitz one.*

For the proof, apply Theorem 3.3.23.

Let us now turn to the regularity problem. Let $\Omega = \mathbb{R}^3_+$ and define the function $\mu(\Gamma)$ in the interval $[0, \Gamma_0]$ as $\mu(\Gamma) = \mu_0 - K\Gamma$, where $K \geqq 0$. Put $\Gamma_0 = \frac{\mu_0}{3K}$. Let us consider the function (6.8.8) in the interval $\left[0, \sqrt{\Gamma_0}\right]$. It is $s' > 0$ in $\left[0, \sqrt{\Gamma_0}\right)$, $s'\left(\sqrt{\Gamma_0}\right) = 0$, and $s''(\gamma) < 0$ in $\left(0, \sqrt{\Gamma_0}\right]$.

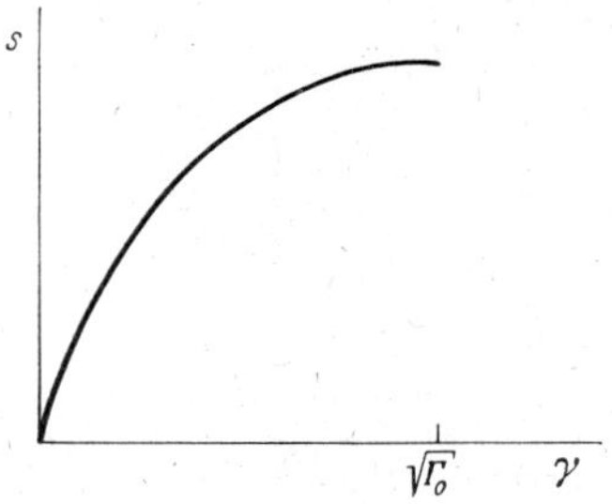

Fig. 6.8.17

This corresponds to mechanical phenomena surprinsingly well. The function $\mu(\gamma^2)$ itself looks like Fig. 6.8.18,

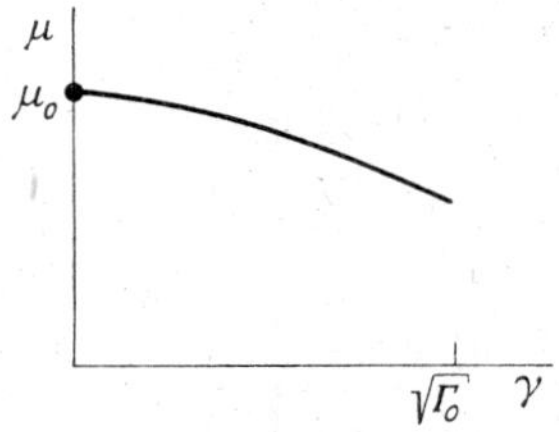

Fig. 6.8.18

which is also reasonable and agrees with experience.

Let $\varepsilon > 0$ and define

$$K_\varepsilon \overset{\text{def}}{=} \{\tilde{u} \in \tilde{W} \mid \|\sqrt{\Gamma(u, u)}\|_{L_\infty(\mathbb{R}^3{}_+)} < \sqrt{\Gamma_0} - \varepsilon\}.$$

Let

(6.8.18)′
$$F(\nabla u) = \frac{k}{2}[e_{ii}(u)]^2 + \frac{1}{2}\int_0^{\Gamma(u)} \mu(t)\,\mathrm{d}t$$

$$= \frac{k}{2}[e_{ii}(u)]^2 + \frac{\mu_0}{2}\Gamma(u, u) - \frac{K}{4}\Gamma^2(u, u),$$

(6.8.19)
$$F_i = \frac{\partial f^i_j}{\partial x_j}, \quad f^i_j \in W^{2,2}(\mathbb{R}^3_+).$$

Let us remark, that the convex of such η, for which $e^*_{ij}e^*_{ij} < \Gamma_0$, if we put $2e_{ij} = \eta^i_j + \eta^j_i$, is not bounded: if we take $\eta^j_i = \delta_{ij}\tau$, then $e^*_{ij}e^*_{ij} = 0$.

6.8.20. THEOREM. *The operator T from Section* 6.6. *defined as* $\left(\tilde{u}_1(.,.,0), \tilde{u}_2(.,.,0), \tilde{u}_3(.,.,0), \frac{\partial F}{\partial \eta_1^1}(\nabla u), \dots \frac{\partial F}{\partial \eta_3^3}(\nabla u)\right)$ *from* $\bar{K}_\varepsilon$ *to* $T(\bar{K})_\varepsilon$ *(recall that* $K_\varepsilon \subset \tilde{W} = [W^{3,2}(\mathbb{R}^3_+)]^3/C$, $T(K_\varepsilon) \subset (\tilde{W}/\tilde{V}) \times [W^{2,2}(\mathbb{R}^3_+)]^3/M$, *see Section* 6.6.) *is a diffeomorphism of* K_ε *onto* $T(K_\varepsilon)$, *a homeomorphism of* $\bar{K}_\varepsilon$ *onto* $T(\bar{K}_\varepsilon) = \overline{T(K_\varepsilon)}$ *and of* ∂K_ε *onto* $\partial T(K_\varepsilon) = T(\partial K_\varepsilon)$, *and* T, T^{-1} *are bounded operators.*

Proof: First, the conditions (6.6.6), (6.6.7) are substituted by

$$\frac{\partial^2 F}{\partial e_{ij}\,\partial e_{kl}}(e)\,\tilde{e}_{ij}\tilde{e}_{kl} \geqq \nu \tilde{e}_{ij}\tilde{e}_{ij}, \tag{6.8.21}$$

$$-\frac{\partial^4 F}{\partial e_{ij}\,\partial e_{kl}\,\partial e_{mn}\,\partial e_{op}}(e)\,\tilde{e}_{ij}\tilde{e}_{kl}\tilde{e}_{mn}\tilde{e}_{op} \geqq 6K[\tilde{e}^*_{ij}\tilde{e}^*_{ij}]^2. \tag{6.8.22}$$

For $\tilde{u} \in \tilde{W}$, we have Korn's inequality

$$\int_{\mathbb{R}^3_+} \frac{\partial u_i}{\partial x_j}\frac{\partial u_i}{\partial x_j}\,dx \leqq c_1 \int_{\mathbb{R}^3_+} e_{ij}(u)\,e_{ij}(u)\,dx \tag{6.8.23}$$

(see, for example, [19]). Actually, let us adapt the proof of Theorem 6.6.16 and of Lemma 6.6.33.

We obtain from (6.6.18) and (6.6.19) that

$$\int_{\mathbb{R}^3_+} f^{i\prime}_j e''_{ij}\,dx = \int_{\mathbb{R}^3_+} \frac{\partial^2 F}{\partial e_{ij}\,\partial e_{kl}} e''_{ij}e''_{kl}\,dx \tag{6.8.24}$$

$$-\frac{1}{3}\int_{\mathbb{R}^3_+} \frac{\partial^4 F}{\partial e_{ij}\,\partial e_{kl}\,\partial e_{mn}\,\partial e_{op}} e'_{ij}e'_{kl}e'_{mn}e'_{po}\,dx$$

$$\geqq \nu \int_{\mathbb{R}^3_+} e''_{ij}e''_{ij}\,dx + 6K\int_{\mathbb{R}^3_+} [e^{*\prime}_{ij}e^{*\prime}_{ij}]^2\,dx.$$

Hence in virtue of (6.8.23), it follows

$$\int_{\mathbb{R}^3_+} \left(\frac{\partial^3 u_i}{\partial x_j\,\partial x_k\,\partial x_l}\right)^2 dx \leqq c_2, \tag{6.8.25}$$

where the value 3 may occur among the subscripts j, k, l at most once.

The equations (6.6.24) have the form

$$\frac{\partial^2 F}{\partial e_{ij}\,\partial e_{kl}}\frac{\partial e_{kl}}{\partial x_j} = \frac{\partial f^i_j}{\partial x_j} \tag{6.8.26}$$

and being differentiated with respect to x_1 or x_2, they imply

$$\frac{\partial^2 F}{\partial e_{ij}\,\partial e_{kl}}\frac{\partial e'_{kl}}{\partial x_j} + \frac{\partial^3 F}{\partial e_{ij}\,\partial e_{kl}\,\partial e^*_{mn}}\frac{\partial e_{kl}}{\partial x_j}e^{*\prime}_{mn} = \frac{\partial f^{i\prime}_j}{\partial x_j}. \tag{6.8.27}$$

Now, the term $\dfrac{\partial^3 F}{\partial e_{ij}\,\partial e_{kl}\,\partial e^*_{mn}}$ is bounded, because it depends on e^*_{op} only.

We have

(6.8.28)
$$\int_{\mathbb{R}^3_+} \left(\frac{\partial e_{kl}}{\partial x_j}\right)^2 (e^{*\prime}_{mn})^2 \,\mathrm{d}x$$
$$\leqq c_2 \left(\int_{\mathbb{R}^3_+} (e^{*\prime}_{mn})^4 \,\mathrm{d}x\right)^{\frac{1}{2}} \cdot (\varepsilon \|u\|'_3 + \lambda(\varepsilon)\,\|u\|_1^{\prime 2}),$$

where the inequality (6.6.41) has been used. But now we can once more use the estimate (6.8.24) (we suppose that $K > 0$, otherwise all is trivial). We thus get from (6.8.27) that

(6.8.29)
$$\int_{\mathbb{R}^3_+} \sum_{i=1}^{3} \left(\frac{\partial^3 u_s}{\partial x_i\,\partial x_3^2}\right)^2 \mathrm{d}x \leqq c_3(\varepsilon) + \varepsilon \|u\|_3^{\prime 2}.$$

Differentiating the system (6.8.26) with respect to x_3, we get an estimate of $\|u\|'_3$. In the proof of Lemma 6.6.33 we use (6.8.23), of course.

6.9. The regularity up to the boundary, conditions of the Liouville type $L(\mathbb{R}^n_+)$

In this section, let us consider, as usual, the system

(6.9.1)
$$-\frac{\partial}{\partial x_i}[a_i^r(x, u, \nabla u)] + a^r(x, u, \nabla u) = f^r, \quad r = 1, 2, \ldots, m,$$

$u = (u_1, u_2, \ldots, u_m)$ in the domain

(6.9.2)
$$\Omega_1 \overset{\text{def}}{=} \{x \in \mathbb{R}^n \mid |x_i| < 1,\ i = 1, 2, \ldots, n,\ x_n > 0\}.$$

Define

(6.9.3)
$$\Omega_t = \{x \in \mathbb{R}^n \mid |x_i| < t,\ i = 1, 2, \ldots, n,\ x_n > 0\},$$
$$\Gamma_t = \{x \in \partial\Omega_t \mid x_n = 0\}.$$

We suppose, as usual, that the functions $a_i^r(x, \xi, \eta)$ and $a^r(x, \xi, \eta)$ are once continuously differentiable with respect to all variables in $\bar{\Omega}_1 \times \mathbb{R}^m \times \mathbb{R}^{mn}$. We assume $f^r \in W^{1,p/2}(\Omega_1)$, $p > n$, and the very strong ellipticity:

(6.9.4)
$$\frac{\partial a_i^r}{\partial \eta_j^s}(x, \xi, \eta)\,\zeta_i^r \zeta_j^s > 0 \quad \text{for } \zeta \neq 0.$$

We suppose from the beginning that $u \in [W^{1,\infty}(\Omega_1)]^m$. Let $u^0 \in [W^{2,p}(\Omega_1)]^m$. Define, as usual,

(6.9.5)
$$V \overset{\text{def}}{=} \{v \in [W^{1,2}(\Omega_1)]^m \mid v_r = 0 \text{ on } \Gamma_1,\ r = 1, 2, \ldots, M,$$
$$0 \leqq M \leqq m,\ v = 0 \text{ on } \partial\Omega_1 \setminus \Gamma_1\}.$$

Further let $h^r(x, \xi)$ be once continuously differentiable functions in $\bar{\Gamma}_1 \times \mathbb{R}^m$ and let $g^r \in W^{1,\infty}(\Gamma_1)$, $r = 1, 2, \ldots, m$.

We shall consider weak solutions such that

(6.9.6) $$u_r - u_r^0 = 0 \quad \text{on } \Gamma_1, \quad r = 1, 2, \ldots, M,$$

(6.9.7) $$\forall v \in V: \int_{\Omega_1} \left[a_i^r(x, u, \nabla u) \frac{\partial v_r}{\partial x_i} + a^r(x, u, \nabla u)\, v_r - f^r v_r \right] \mathrm{d}x + \int_{\Gamma_1} [h^r(x, u) - g^r]\, v_r \, \mathrm{d}x' = 0.$$

The integration by parts in a formal way gives the system (6.9.1) and the boundary conditions (unstable)

(6.9.8) $$-a_n^r(x, u, \nabla u) + h^r(x, u) - g^r(x) = 0 \quad \text{on } \Gamma_1, \quad r = M + 1, \ldots, m.$$

6.9.9. DEFINITION. *The solutions to* (6.9.6), (6.9.7) *are called* ***regular up to the boundary*** Γ_1, *provided that for every* $t < 1$ *and for the family of solutions such that*

(6.9.10) $$\|u\|_{[W^{1,\infty}(\Omega_1)]^m} + \|f\|_{[W^{1,p/2}(\Omega_1)]^m} + \|u^0\|_{[W^{2,p}(\Omega_1)]^m} + \|g\|_{[W^{1,\infty}(\Gamma_1)]^m} \leqq \mu < \infty,$$

there exists $c(t, \mu)$ *such that for* $\alpha = 1 - \dfrac{n}{p}$ *it is*

(6.9.11) $$\|u\|_{[C^{1,\alpha}(\bar{\Omega}_t)]^m} \leqq c(t, \mu).$$

6.9.12. DEFINITION. *Consider* $\phi \in \mathscr{D}(\mathbb{R}^n)$ *such that* $\phi_r = 0$ *for* $x_n = 0$, $r = 1, 2, \ldots, M$. *Let* $x^0 \in \Gamma_1$, $\xi \in \mathbb{R}^m$, $d \in \mathbb{R}^m$, p_r, $r = 1, 2, \ldots, M$, *be polynomials of at most first degree in* $\mathbb{R}^{n-1}$. *Consider* v *in* $\mathbb{R}_+^n$ *such that for* $x_n = 0$, $v_r = p_r$, $r = 1, 2, \ldots, M$, *are the solutions to the system*

(6.9.13) $$\int_{\mathbb{R}_+^n} a_i^r(x^0, \xi, \nabla v) \frac{\partial \phi_r}{\partial x_i} \, \mathrm{d}x = \int_{\mathbb{R}^{n-1}} d^r \phi_r \, \mathrm{d}x'.$$

Then the problem (6.9.6), (6.9.7) *satisfies the condition* $L(\mathbb{R}_+^n)$ *if* $\forall x^0 \in \Gamma_1$, $\forall \xi \in \mathbb{R}^m$, *the solution* v *in question is a polynomial of at most first degree.*

Let us begin with the easy part of the regularity, with the "necessity" of the condition $L(\mathbb{R}_+^n)$.

6.9.14. THEOREM. *In* Ω_1, *consider the family of solutions to the problem*

(6.9.15) $$u_r - u_r^0 = 0 \quad \textit{on } \Gamma_1, \qquad r = 1, 2, \ldots, M,$$

(6.9.16) $$\forall v \in V: \int_{\Omega_1} a_i^r(x^0, \xi, \nabla u) \frac{\partial v_r}{\partial x_i} \, \mathrm{d}x = \int_{\Gamma_1} g^r v_r \, \mathrm{d}x', \quad x^0 \in \Gamma_1, \quad \xi \in \mathbb{R}^m,$$

(6.9.17) $$\|u\|_{[W^{1,\infty}(\Omega_1)]^m} + \|u^0\|_{[W^{2,p}(\Omega_1)]^m} + \|g\|_{[W^{1,\infty}(\Gamma_1)]^m} \leqq \mu.$$

Suppose that with some $\alpha > 0$ the inequality

(6.9.18)
$$\|u\|_{[C^{1,\alpha}(\Omega_{\frac{1}{2}})]^m} \leqq c(\mu)$$

holds.

Then for x^0 and ξ in question, the condition $L(\mathbb{R}^n_+)$ is satisfied.

Proof. Let v be a solution to (6.9.13) and put $v^R(y) = \dfrac{1}{R} v(Ry)$. For R large enough, v^R is from the family (6.9.16), (6.9.17). But in $\Omega_{1/2}$ it is, with $y = \dfrac{x}{R}$,

(6.9.19)
$$\left|\frac{\partial v^R}{\partial y_i}(y) - \frac{\partial v^R}{\partial y_i}(0)\right| = \left|\frac{\partial v}{\partial x_i}(x) - \frac{\partial v}{\partial x_i}(0)\right| \leqq c\left|\frac{x}{R}\right|^\alpha$$

for $R \geqq 2|x|$. Hence we get the result as $R \to \infty$.

Let us consider now the sets

$$U_t(x^0) = \{x \mid |x_i - x_i^0| < t, \quad i = 1, 2, \ldots, n, \ x_n > 0, \ x_n^0 \geqq 0\}, \quad x^0 \in \Omega_1.$$

Let b_{ij}^{rs} be constant coefficients such that

(6.9.21)
$$b_{ij}^{rs}\zeta_i^r\zeta_j^s \geqq \nu|\zeta|^2, \quad \nu > 0,$$

and let $\phi \in [\mathscr{D}(B_1(x^0))]^{m'}$ be such that $\phi_r = 0$ for $x_n = 0$, $r = 1, 2, \ldots, M'$, $0 \leqq M' \leqq m'$, if $t \geqq x_n^0$. Let $v \in [W^{1,2}(U_1(x^0))]^{m'}$ be the weak solutions to the problem

(6.9.22)
$$v_r = 0 \quad \text{for} \quad x_n = 0, \quad r = 1, 2, \ldots, M', \quad (\text{if } t \geqq x_n^0)$$

and such that $\forall\phi$ in question,

(6.9.23)
$$\int_{U_1(x^0)} b_{ij}^{rs} \frac{\partial v_s}{\partial x_j} \frac{\partial \phi_r}{\partial x_i} \, dx = 0.$$

We have

6.9.24. LEMMA. $t^{-n+2} \displaystyle\int_{U_t(x^0)} |\nabla v|^2 \, dx \leqq kt^2 \int_{U_1(x^0)} |\nabla v|^2 \, dx,$ *where* $k = k(\nu, \max |b_{ij}^{rs}|)$.

Proof. Consider the more difficult case of $t \geqq x_n^0$. We have by virtue of Theorem 5.2.30, which can be repeated, with regard to the constant coefficients, arbitrarily many times, and by virtue of the imbedding $W^{k,2}(U_{\frac{1}{2}}(x^0)) \hookrightarrow C(\overline{U_{\frac{1}{2}}(x^0)})$, $2k > n$, that

(6.9.25)
$$\max_{U_{\frac{1}{2}}(x^0)} |\nabla v(x)|^2 \leqq c_1 \int_{U_1(x^0)} |\nabla v|^2 \, dx.$$

Then the result follows.

In order to avoid some technical difficulties, we shall in the sequel consider only the systems

(6.9.26)
$$\int_\Omega \left[a_i^r(\nabla u)\frac{\partial v_r}{\partial x_i} - f^r v_r\right] dx + \int_{\Gamma_1} [h^r(u) - g^r] \, v_r \, dx' = 0,$$

$\forall v \in V$; clearly (6.9.8) is preserved.

For the general systems, see the work [72].

Put $\omega_r^\alpha = \dfrac{\partial u_r}{\partial x_\alpha}$, $r = 1, 2, \ldots, m$, $\alpha = 1, 2, \ldots, n-1$. For ω_r^α we have the system in variations

$$(6.9.27)\qquad \int\limits_{U_t(x^0)} \left[\frac{\partial a_i^r}{\partial \eta_j^s}\frac{\partial \omega_s^\alpha}{\partial x_j}\frac{\partial \phi_r^\beta}{\partial x_i}\delta_{\alpha\beta} - \frac{\partial f^r}{\partial x_\alpha}\phi_\alpha^r\right] \mathrm{d}x + \int\limits_{\Gamma_t(x^0)} \left[\frac{\partial h^r}{\partial u_s}\omega_s^a - \frac{\partial g^r}{\partial x_\alpha}\right]\phi_r^a \, \mathrm{d}x' = 0,$$

where $\phi_r^\alpha = 0$, $r = 1, 2, \ldots, M$, $\alpha = 1, 2, \ldots, n-1$, on $\Gamma_t(x^0) \stackrel{\text{def}}{=} \overline{U_t(x^0)} \cap \{x \mid x_n = 0\}$, $\phi_\alpha^r = 0$ for all the indices on $\partial U_t(x^0) \setminus \Gamma_t(x^0)$.

6.9.28. DEFINITION. *Let w be the solution to the system*

$$(6.9.29)\qquad \int\limits_{U_t(x^0)} \frac{\partial a_i^r}{\partial \eta_j^s}\frac{\partial w_s^\alpha}{\partial x_j}\frac{\partial \phi_r^\beta}{\partial x_i}\delta_{\alpha\beta} = \int\limits_{U_t(x^0)} \frac{\partial f^r}{\partial x_\alpha}\phi_r^\alpha \, \mathrm{d}x + \int\limits_{\Gamma_t(x^0)} \left[\frac{\partial g^r}{\partial x_\alpha} - \frac{\partial h^r}{\partial u_s}\omega_s^a\right]\phi_r \, \mathrm{d}x'$$

with the boundary conditions

$$(6.9.30)\qquad w_r^a = \frac{\partial u_r^0}{\partial x_\alpha}, \quad r = 1, 2, \ldots, M, \quad \alpha = 1, 2, \ldots, n-1,$$

on $\Gamma_t(x^0)$,

$$(6.9.30)'\qquad w_r^\alpha = \frac{\partial u_r^0}{\partial x_\alpha}, \quad r = 1, 2, \ldots, m, \quad \alpha = 1, 2, \ldots, n-1,$$

on $\partial U_t(x^0) \setminus \Gamma_t(x^0)$.

Decompose $\omega = v + w$.

6.9.31. LEMMA. *For every $0 < \tau < 1$ and the family of solutions* (6.9.10), *there exist $\varepsilon_0 = \varepsilon_0(\tau, \mu, t)$, $R_0 = R_0(\tau, \mu, t)$ for $0 < t < 1$ such that if $R \leqq R_0$, $x^0 \in \bar{\Omega}_t$, and $R^{-n+2} \int_{U_R(x^0)} |\nabla v|^2 \, \mathrm{d}x < \varepsilon_0^2$, then*

$$(6.9.32)\qquad (R\tau)^{-n+2} \int\limits_{U_{R\tau}(x^0)} |\nabla v|^2 \, \mathrm{d}x \leqq 2K\tau^2 R^{-n+2} \int\limits_{U_R(x^0)} |\nabla v|^2 \, \mathrm{d}x,$$

where K is the coefficient from Lemma 6.9.24 *for the family of the coefficients $\dfrac{\partial a_i^r}{\partial \eta_j^s}(\eta)\,\delta_{\alpha\beta}$, $|\eta| \leqq \mu$.*

Proof. Let us suppose the contrary. Then there exist $\varepsilon_\nu \to 0$, $R_\nu \to 0$, $x_\nu \to x^0 \in \bar{\Omega}_t$, v^ν generated by our family (this means that we come from u^ν to ${}^\nu\omega$ and then we decompose ${}^\nu\omega$ according to Definition 6.9.28) and such that

$$R_\nu^{-n+2} \int\limits_{U_{R_\nu}(x^\nu)} |\nabla v^\nu|^2 \, \mathrm{d}x = \varepsilon_\nu^2$$

and

(6.9.33) $$\tau^{-n+2} R_\nu^{-n+2} \int_{U_{R_\nu\tau}(x^\nu)} |\nabla v^\nu|^2 \, dx > 2K\tau^2 R_\nu^{-n+2} \int_{U_{R_\nu}(x^\nu)} |\nabla v^\nu|^2 \, dx.$$

We have as in Section 6.2. that

(6.9.34) $$\int_{U_{R_\nu}(x^\nu)} |\nabla w^\nu|^2 \, dx \leqq c_1 R_\nu^{n\left(1-\frac{2}{p}\right)},$$

hence

(6.9.35) $$\int_{U_{R_\nu}(x^\nu)} |\nabla \omega^\nu|^2 \, dx \leqq c_2 \left[R_\nu^{n\left(1-\frac{2}{p}\right)} + \varepsilon_\nu^2 R^{n-2}\right].$$

But from the system

(6.9.36) $$-\frac{\partial}{\partial x_i}[a_i^r(\nabla u)] = f^r, \quad r = 1, 2, \ldots, m,$$

and from (6.9.35), we can calculate the estimate

(6.9.37) $$\int_{U_{R_\nu}(x^\nu)} \frac{\partial^2 u_r}{\partial x_n^2} \frac{\partial^2 u_r}{\partial x_n^2} \, dx \leqq c_3 \left[R_\nu^{n\left(1-\frac{2}{p}\right)} + \varepsilon_\nu^2 R^{n-2}\right].$$

Put $x = x^\nu + R_\nu y$, $\tilde{v}^\nu(y) = \dfrac{1}{\varepsilon_\nu}[v^\nu(x^\nu + R_\nu y) - \varkappa^\nu]$, $\tilde{w}^\nu(y) = w^\nu(x^\nu + R_\nu y)$, where

(6.9.38) $$\varkappa^\nu = \frac{1}{|U_{R_\nu}(x^\nu)|} \int_{U_{R_\nu}(x^\nu)} v^\nu \, dx$$

if $R_\nu < x_n^\nu$. If such a subsequence does not exist (where $R_\nu < x_n^\nu$), put

(6.9.39) $$\varkappa^\nu = \frac{1}{|\Gamma_{R_\nu}(x^\nu)|} \int_{\Gamma_{R_\nu}(x^\nu)} v^\nu \, dx'.$$

We shall study the latter, more difficult case. We have, with a clear notation, that

(6.9.40) $$1 = R_\nu^{-n+2} \varepsilon_\nu^{-2} \int_{U_{R_\nu}(x^\nu)} |\nabla v^\nu|^2 \, dx = \int_{U^\nu} |\nabla \tilde{v}^\nu|^2 \, dy,$$

(6.9.41) $$\varepsilon_\nu^{-2} \tau^{-n+2} R_\nu^{-n+2} \int_{U_{\tau R_\nu}(x^\nu)} |\nabla v^\nu|^2 \, dx = \tau^{-n+2} \int_{U_\tau^\nu} |\nabla \tilde{v}^\nu|^2 \, dy.$$

On Γ_ν, the part of ∂U^ν, corresponding to $\Gamma_{R_\nu}(x^\nu)$, we have

(6.9.42) $$0 = \int_{\Gamma_\nu} \tilde{v}^\nu \, dy',$$

hence

(6.9.43) $$\int_{U^\nu} |\tilde{v}^\nu|^2 \, dy \leqq c_4 \int_{U^\nu} |\nabla \tilde{v}^\nu|^2 \, dy.$$

Now

(6.9.44) $$\omega^\nu(x^\nu + R_\nu y) = \tilde{v}^\nu(y)\, \varepsilon_\nu + \tilde{w}^\nu(y) + \varkappa^\nu.$$

But

$$\int_{U_\nu} |\nabla_y \tilde{w}^\nu|^2 \, dy = R_\nu^{2-n} \int_{U_{R_\nu}(x^\nu)} |\nabla w^\nu|^2 \, dx \leq c_1 R_\nu^{2\left(1-\frac{n}{p}\right)}. \tag{6.9.44$'$}$$

We have the inequality

$$\int_{U_\nu} \left| \tilde{w}_r^{\nu\alpha} - \frac{\partial u_r^0}{\partial x_\alpha}(x^\nu + R_\nu y) \right|^2 dy \leq c_5 \int_{U_\nu} \left| \nabla_y \tilde{w}_r^{\nu\alpha} - \nabla_y \frac{\partial u_r^0}{\partial x_\alpha} \right|^2 dy, \tag{6.9.45}$$

$$r = 1, 2, \ldots, m, \quad \alpha = 1, 2, \ldots, n-1,$$

by virtue of thc boundary conditions on $\partial U_{R_\nu}(x^\nu) \setminus \Gamma_{R_\nu}(x^\nu)$. In order to avoid technical complications let us suppose that $x_n^\nu = 0$. We have $U^\nu = \Omega_1$ and it follows from (6.9.45) that $\tilde{w} \to \tilde{c}$, a constant function almost everywhere in Ω_1. We can, however, suppose in virtue of (6.9.43) that $\varepsilon_\nu \tilde{v}^\nu \to 0$ in Ω_1 almost everywhere. But $\omega^\nu(x^\nu + R_\nu y)$ are bounded in Ω_1, hence we can suppose that $\varkappa^\nu \to \varkappa \in \mathbb{R}$. Hence $\omega^\nu(x^\nu + R_\nu y) \to \tilde{\varkappa}$ and $|\tilde{\varkappa}| \leq \mu$ almost everywhere in Ω_1. Now

$$R_\nu^{-2} \int_{U_{R_\nu}(x^\nu)} \left| \frac{\partial u^\nu}{\partial x_n} - \frac{1}{|U_{R_\nu}(x^\nu)|} \int_{U_{R_\nu}(x^\nu)} \frac{\partial u^\nu}{\partial x_n} \, dx \right|^2 dx \tag{6.9.46}$$

$$\leq c_6 \int_{U_{R_\nu}(x^\nu)} \left| \nabla \frac{\partial u^\nu}{\partial x_n} \right|^2 dx,$$

thus we can suppose that $\dfrac{\partial u^\nu}{\partial x_n}(x^\nu + R_\nu y) \to \tilde{v}^n \in \mathbb{R}^m$ almost everywhere in Ω_1 and $|\tilde{v}| \leq \mu$. The function $\tilde{v}^\nu$ satisfies the system

$$\int_\Omega \frac{\partial a_i^r}{\partial \eta_j^s}(\nabla u^\nu(x^\nu + R_\nu y)) \frac{\partial \tilde{v}_s^{\nu\alpha}}{\partial y_j} \frac{\partial \psi_r^\beta}{\partial y_i} \delta_{\alpha\beta} \, dy = 0 \tag{6.9.47}$$

with $\tilde{v}_s^{\nu\alpha} = 0$ on Γ_1 for $\alpha = 1, 2, \ldots, n-1$, $s = 1, 2, \ldots, M$, $\psi_r^\beta = 0$ on Γ_1 for the same indices, and $\psi = 0$ on $\partial\Omega_1 \setminus \Gamma_1$. But $\dfrac{\partial a_i^r}{\partial \eta_j^s}(\nabla u^\nu(x^\nu + R_\nu y)) \to \dfrac{\partial a_i^r}{\partial \eta_j^s}(\tilde{\varkappa}, \tilde{v})$ almost everywhere in Ω_1. It is a standard procedure to extract a subsequence from $\tilde{v}^\nu$ (denoting it by the same symbol) in such a way that $\tilde{v}^\nu \rightharpoonup \tilde{v}$ (weakly) in $[W^{1,2}(\Omega_1)]^{m(n-1)}$ and strongly in $[W^{1,2}(\Omega_t)]^{m(n-1)}$ for every $t < 1$. From this it follows that the function $\tilde{v}$ is a solution to the system (6.9.47) with the limit constant coefficients. From (6.9.33), (6.9.40) and (6.9.41) the inequality $\tau^{-n+2} \int_{\Omega_\tau} |\nabla_y \tilde{v}|^2 \, dy \geq 2K\tau^2 > 0$, follows, so $\tilde{v} \neq 0$ and, a fortiori, since $\int_{\Omega_1} |\nabla_y \tilde{v}|^2 \, dy \leq 1$ (see (6.9.40)), we have also the inequality $\tau^{-n+2} \int_{\Omega_\tau} |\nabla_y v|^2 \, dy \geq 2K\tau^2 \int_{\Omega_1} |\nabla_y v|^2 \, dy$, which contradicts Lemma 6.9.24.

6.9.48. LEMMA. *For the family* (6.9.10) *of solutions to the problem* (6.9.6), (6.9.7) *and for* $x^0 \in \bar{\Omega}_t$, $t < 1$, *in the notation of Lemma* 6.9.28, *it is*

$$\lim_{R\to 0} R^{-n+2} \int_{U_R(x^0)} |\nabla v|^2 \, dx = 0$$

uniformly with respect to u from the family mentioned and $x^0 \in \bar{\Omega}_t$, *provided that* $L(\mathbb{R}^n)$ *and* $L(\mathbb{R}^n_+)$ *are satisfied.*

Proof. Let us assume the contrary. So we have a sequence $x^\nu \to x^0$ in $\bar{\Omega}_t$, $R_\nu \to 0$, and $R_\nu^{-n+2} \int_{U_{R_\nu}(x^\nu)} |\nabla v^\nu|^2 \, dx \geqq \varepsilon > 0$. Use the transformation $x = x^\nu + R_\nu y$ which maps Ω_1 onto Ω^ν. If, after the extraction of a subsequence, the relation $\dfrac{x_n^\nu}{R_\nu} \to \infty$ takes place, the whole problem leads to the use of $L(\mathbb{R}^n)$, because $\Omega^\nu \to \mathbb{R}^n$ in this case. We shall consider the more complicated situation, i.e. that $\dfrac{x_n^\nu}{R_\nu} \to a \geqq 0$. We have $\Omega^\nu \to \mathbb{R}^n_+$. Apart from some technical complications, the proof proceeds as in the case of $x_n^\nu = 0$. So let us suppose this. Let us introduce $\tilde{u}^\nu(y) = \dfrac{1}{R_\nu}[u^\nu(x^\nu + R_\nu y) - u^\nu(x^\nu)]$. We obtain as in the proof of Theorem 6.3.8 that, for every $B_a(O) \cap \Omega^\nu$:

$$(6.9.49) \qquad \int_{B_a(O)\cap\Omega^\nu} \sum_{i,j,r} \left(\frac{\partial^2 \tilde{u}_r^\nu}{\partial y_i \, \partial y_j}\right)^2 dy \leqq c(a).$$

Since

$$(6.9.50) \qquad \|\tilde{u}^\nu\|_{[W^{1,\infty}(\Omega^\nu)]^m} \leqq c_1,$$

we can suppose $\tilde{u}^\nu \to \tilde{u}$ in $[W^{1,2}(B_a(O) \cap \mathbb{R}^n_+)]^m$ and in $[W^{1,2}(B_a(O) \cap \partial\mathbb{R}^n_+)]^m$ $\forall a > 0$. Now

$$(6.9.51) \qquad \int_{\mathbb{R}^n_+} \left[a_i^r(\nabla_y \tilde{u}^\nu) \frac{\partial \phi_r}{\partial y_i} - R_\nu f_r(x^\nu + R_\nu y)\right] dy$$

$$= \int_{\mathbb{R}^{n-1}} [h^r(\tilde{u}^\nu(y) R_\nu + u^\nu(x^\nu)) - g^r(x^\nu + R_\nu y)] \, \phi_r \, dy'.$$

But $u_r^{0\,\nu}(y) = \dfrac{1}{R_\nu}[u_r^0(x^\nu + R_\nu y) - u_r^0(x^\nu)] \to p_r(y)$, polynomials of at most first degree, $r = 1, 2, \ldots, M$. Hence $\tilde{u}$ in $\mathbb{R}^n_+$, which clearly satisfies $|\nabla_y \tilde{u}| \leqq c < \infty$, is a polynomial of at most first degree in virtue of the condition $L(\mathbb{R}^n_+)$. Actually, it is necessary to consider the system (6.9.27) for $\tilde{\omega}^\nu$ and the domain $B_a(O) \cap \mathbb{R}^n_+$. For the test functions we here take $\phi_r^{\nu\alpha} = \Big\{\tilde{\omega}_r^{\nu\alpha} - \dfrac{\partial u^0}{\partial x_\alpha}(x^\nu + R_\nu y) - \dfrac{1}{|\Gamma_a|}\displaystyle\int_{\Gamma_a}\Big[\tilde{\omega}_r^{\nu\alpha} - \dfrac{\partial u^0}{\partial x_\alpha}(x^\nu + R_\nu y)\Big] dy'\Big\} \, \psi_r(y)$, $r = 1, 2, \ldots, m$, $\alpha = 1, 2, \ldots, n-1$, where $\Gamma_a = B_a(O) \cap \partial\mathbb{R}^n_+$ and $\psi \in \mathscr{D}(B_a(O))$, $\psi(y) = 1$ in $B_{a/2}(O)$. But we have $\lim_{\nu\to\infty} \int_{B_a(O)\cap\mathbb{R}^n_+} |\phi^\nu|^2 \, dy = 0$, which gives

$\lim_{\nu \to \infty} \int_{B_1(O) \cap \mathbb{R}^n_+} |\nabla \tilde{\omega}|^2 \, dy = \lim_{\nu \to \infty} R_\nu^{-n+2} \int_{U_{R_\nu}(x^\nu)} |\nabla \omega|^2 \, dx = 0$. In virtue of (6.9.44)′, we get a contradiction.

6.9.52. THEOREM. *Let us consider the problem* (6.9.6), (6.9.7) *and let us suppose that the conditions* $L(\mathbb{R}^n)$, $L(\mathbb{R}^n_+)$ *are satisfied. Hence the regularity in the sense of Definition* 6.9.9 *holds.*

Idea of the proof: We use Lemmas 6.9.31, 6.9.48 and an analogue to Lemma 6.2.25 where $R^{-n+2} \int_{U_R(x^0)} |\nabla v|^2 \, dx$ has to be considered. We combine the inequalities (6.9.32) and (6.9.34) and we come finally to

$$R^{-n+2} \int_{U_R(x^0)} \frac{\partial^2 u_r}{\partial x_i \, \partial x_j} \frac{\partial^2 u_r}{\partial x_i \, \partial x_j} \, dx \leq c R^{2\left(1 - \frac{n}{p}\right)}, \tag{6.9.53}$$

which immediately implies (6.9.11), see the proof of Theorem 2.2.13.

References

[1] Ladyženskaja, O. A.; Ural'ceva, N. N.: *Lineĭnye i kvazilineĭnye uravnenija elliptičeskogo tipa.* Nauka, Moskva 1973.

[2] Morrey, Ch. B.: *Multiple integrals in the calculus of variations.* Springer, 1966.

[3] Lions, J. L.: *Quelques méthodes de résolution des problèmes aux limites non linéaires.* Dunod, Paris 1969.

[4] Gilbarg, D.; Trudinger, N. S.: *Elliptic Partial Differential Equations of Second Order.* Springer, 1977.

[5] Fučík, S.; Kufner, A.: *Nonlinear differential equations.* Elsevier, Scient. Publ. Comp., Amsterdam–Oxford–New York 1980.

[6] Giusti, E.: *Equazioni ellittiche del secondo ordine.* Pitagora Editrice, Bologna 1978.

[7] Fučík, S.; Nečas, J.; Souček, V.: *Einführung in die Variationsrechnung.* TEUBNER-TEXTE zur Mathematik 11, Leipzig 1977.

[8] Zeidler, E.: *Vorlesungen über nichtlineare Funktionalanalysis I–III.* TEUBNER-TEXTE zur Mathematik 2, 9, 16, Leipzig 1976–1978.

[9] Sobolev, S. L.: *Nekotorye priloženija funkcional'nogo analiza k matematičeskoĭ fizike.* LGU, 1950.

[10] Vajnberg, M. M.: *Variacionnye metody issledovanija nelineĭnyh operatorov.* GITTL, Moskva 1956.

[11] Courant, R.; Hilbert, D.: *Methoden der Mathematischen Physik.* Berlin, Springer, 1937.

[12] Céa, J.: *Optimisation, théorie et algorithmes.* Dunod, Paris 1971.

[13] Morrey, Ch. B.: *Existence and differentiability theorems for the solutions of variational problems for multiple integrals.* Bull. Amer. Math. Soc. 46 (1940), 439–458.

[14] de Giorgi, E.: *Sulla differenziabilità e analiticità delle estremali degli integrali multipli regolari.* Mem. Accad. Sci. Torino, Cl. Sci. Fis. Mat. Nat. 3 (1957), 25–43.

[15] Giaquinta, M.; Nečas, J.: *On the regularity of weak solutions to nonlinear elliptic systems via Liouville's type property.* Comm. Math. Univ. Carol. 20 (1979), 111–122.

[16] GIAQUINTA, M.; NEČAS, J.: *On the regularity of weak solutions to nonlinear elliptic systems of partial differential equations.* Journ. reine und angew. Math., Band 316 (1980), 140–159.

[17] GIUSTI, E.: *Minimal surfaces and functions of bounded variation.* Notes on Pure Mathematics 10, 1977, Depart. of Math., Austr. Nat. University, Canberra.

[18] WASHIZU, K.: *Variational Methods in Elasticity and Plasticity.* Pergamon Press, 1968.

[19] NEČAS, J.; HLAVÁČEK, I.: *Mathematical Theory of Elastic and Elasto-Plastic Bodies: An Introduction.* Elsevier, SNTL, 1981.

[20] BALL, J.: *Convexity Conditions and Existence Theorems in Nonlinear Elasticity.* Archive for Rational Mech. Anal. Vol. 63, N. 4 (1977), 337–403.

[21] NEČAS, J.: *On the regularity of weak solutions to nonlinear elliptic systems of partial differential equations.* Lectures 1979, Scuola Normale Sup. Pisa.

[22] NEČAS, J.: *Régularité des solutions faibles aux équations elliptiques non-linéaires avec applications à elasticité.* Analyse Numérique, Université Pierre et Marie Curie, 1981.

[23] TRIEBEL, H.: *Fourier analysis and function spaces.* TEUBNER-TEXTE zur Mathematik 7, Leipzig 1977.

[24] TRIEBEL, H.: *Spaces of Besov-Hardy-Sobolev type.* TEUBNER-TEXTE zur Mathematik 15, Leipzig 1978.

[25] NEČAS, J.: *Les méthodes directes en théorie des équations elliptiques.* Academia, Prague 1967.

[26] KUFNER, A.; JOHN, O.; FUČÍK, S.: *Function spaces.* Academia, Prague 1977.

[27] BESOV, O. V.; IL'IN, V. P.; NIKOLSKIĬ, S. M.: *Integral'nye predstavlenija funkciĭ i teoremy vloženija.* Nauka, Moskva 1975.

[28] ADAMS, R. A.: *Sobolev spaces.* Academic Press, 1975.

[29] PASCALI, D.; SBURLAN, S.: *Nonlinear mappings of monotone type.* Editura Academiei, Bucuresti 1978.

[30] BRÉZIS, H.: *Opérateurs maximaux monotones.* North-Holland 1973.

[31] GAJEWSKI, H.; GRÖGER, K.; ZACHARIAS, K.; *Nichtlineare Operatorgleichungen und Operatordifferentialgleichungen.* Akademie-Verlag, Berlin 1974.

[32] FUČÍK, S.; NEČAS, J.; SOUČEK, J.; SOUČEK, V.: *Spectral analysis of nonlinear operators.* Springer, 1973.

[33] SKRYPNIK, I. V.: *Nelineĭnye elliptičeskie uravnenija vysšego porjadka.* Kiev 1973.

[34] BRÉZIS, H.; NIRENBERG, L.: *Characterisations of the range of some nonlinear operators and applications to boundary value problems.* Annali Scuola Norm. Sup. Pisa, 1978, 818–919.

[35] GOSSEZ, J. P.: *Nonlinear Elliptic Boundary Value Problems for Equations with Rapidly or Slowly Increasing Coefficients.* Trans. Amer. Math. Soc. 190 (1974), 163–205.

[36] Giaquinta, M.: *Multiple integrals in the calculus of variations and non linear elliptic systems*. Universität Bonn, preprint no. 443, 1981, Sonderforschungsbereich 72.

[37] Landesman, E. M.; Lazer, A. C.: *Nonlinear Perturbations of Linear Elliptic Boundary Value Problems at Resonance*. J. Math. Mech. 19 (1970), 609–623.

[38] Nečas, J.: *On the Ranges of Nonlinear Operators with Linear Asymptotes which are not Invertible*. Comment. Math. Univ. Carol. 14 (1973), 63–72.

[39] Fučík, S.: *Solvability of Nonlinear Equations and Boundary Value Problems*. Society of Czechoslovak mathematicians and physicists, Prague 1980.

[40] Nečas, J.: *Les équations elliptiques non linéaires*. Czechoslovak Math. J. 19 (1969), 252–274.

[41] Jarušek, J.; Nečas, J.: *Sur les domaines des valeurs des opérateurs nonlinéaires*. Čas. Pěst. Mat. 102 (1977), 61–72.

[42] Schwetlick, H.: *Numerische Lösung nichtlinearer Gleichungen*. VEB Deutscher Verlag der Wissenschaften, Berlin 1979.

[43] Kantorovič, L. V.; Akilov, G. P.: *Funkcional'nyĭ analiz*. Nauka, Moskva 1977.

[44] Meyers, N. G.: *An L^p-estimate for the gradient of solutions of second order elliptic divergence equations*. Ann. S.N.S. Pisa (3), 17 (1963), 189–206.

[45] Nečas, J.: *Sur la régularité des solutions faibles des équations elliptiques non linéaires*. Comm. Math. Univ. Carolinae 9, 3 (1968), 365–414.

[46] Nečas, J.: *Variační metody v nelineární pružnosti*. Acta polytechnica, ČVUT, Praha 1973, 129–133.

[47] Stampacchia, G.: *Equations elliptiques du second order à coefficients discontinues*. Les Presses de l'Univ. de Montréal (1966).

[48] Nečas, J.: *Ob oblastjah tipa* $\mathfrak{N}$. Czechoslovak math. J. 12 (1962), 277–287.

[49] Dubinskiĭ, Ju. A.: *Nonlinear elliptic and parabolic Equations*. Itogi Nauki i Techniki, Sovremennye Problemy Matematiki, 9 (1976), Engl. Transl. J. of Soviet Math. 12 (5), 475–554 (1979).

[50] Müller, M.; Nečas, J.: *Über die Regularität der schwachen Lösungen von Randwertaufgaben für quasilineare elliptische Differentialgleichungen höherer Ordnung*. Czechoslovak Math. J. 25 (100), 1975, 227–239.

[51] Nečas, J.; Stará, J.: *Principio di massimo per i sistemi ellittici quasilineari non diagonali*. Boll. UMI 6 (1972).

[52] Moser, J.: *A new proof of De Giorgi's theorem concerning the regularity problem for elliptic differential equations*. Comm. Pure Appl. Math. 13, (1960), 457–468.

[53] Moser, J.: *An Harnack's theorem for elliptic differential equations*. Comm. Pure Appl. Math. 14 (1961), 577–591.

[54] Stampacchia, G.: *On some Regular Multiple Integral Problems in the Calculus of Variations*. Comm. Pure Appl. Math. vol. 16 (1963), 383–421.

[55] Hartman, P.; Stampacchia, G.: *On some nonlinear elliptic differential-functional equations*. Acta Math. vol. 115, 1966.

[56] HARTMAN, P.: *On the bounded slope condition.* Pacific J. Math.

[57] NASH, J.: *Continuity of the solutions of parabolic and elliptic equations.* Ann. J. Math. 80 (1958).

[58] DANĚČEK, J.: *Thesis.* 1982.

[59] AGMON, S.; DOUGLIS, A.; NIRENBERG, L.: *Estimates near the boundary for solutions of elliptic partial differential equations satisfying general boundary conditions I, II.* Comm. Pure Appl. Math. 12 (1959), 623–727, 17 (1964), 35–92.

[60] DE GIORGI, E.: *Un esempio di estremali discontinue per un problema variazionale di tipo ellitico.* Boll. UMI 4 (1968).

[61] GIUSTI, E.; MIRANDA, M.: *Un esempio di soluzioni discontinue per un problema di minima relativo ad un integrale regolare del calcolo delle variationi.* Boll. UMI 2, 1–8 (1968).

[62] KAWOHL, B.: *On Liouville theorem, continuity and Hölder continuity of weak solutions to some quasilinear elliptic systems.* Comm. Math. Univ. Carol. (1980).

[63] GIUSTI, E.: *Regolarità parziale delle soluzioni ai sistemi ellittici quasi lineari di ordine arbitrario.* Ann. S.N.S. Pisa 23 (1969), 115–141.

[63′] MORREY, CH. B.: *Partial regularity results for non linear elliptic systems.* Journ. Math. Mech. 17 (1968), 649–670.

[64] NEČAS, J.; JOHN, O.; STARÁ, J.: *Counterexample to the regularity of weak solution of elliptic system.* Comm. Math. Univ. Carol. 21, 1 (1980), 145–154.

[65] NEČAS, J.: *A necessary and sufficient condition for the regularity of weak solutions to nonlinear elliptic systems of partial differential equations.* Abhandlungen der Akad. Wissen. DDR, 1981.

[66] NEČAS, J.: *Example of an irregular solution to a nonlinear elliptic system with analytic coefficients and conditions for regularity.* Theory of Non Linear Operators. Abhandl. Akad. Wissen. DDR, 1977.

[67] NEČAS, J.: *On the regularity of weak solutions to variational equations and inequalities for nonlinear second order elliptic systems.* Equadiff IV, Praha, Springer-Verlag, Lecture Notes 703.

[68] KOŠELEV, S. I.: *Regularity of solutions of quasilinear elliptic systems.* Uspehi Mat. Nauk 13: 4, B. 49 (1978), Engl. Transl. Russ. Math. Surv. 33: 4 (1978), 1–52.

[69] NEČAS, J.; OLEĬNIK, O. A.: *Teoremy Liuvillja dlja elliptičeskih sistem.* Dokl. Akad. Nauk, SSSR, 1980, t. 252, n. 6, 1312–1316.

[70] NEČAS, J.: *Régularité des solutions faibles aux équations elliptiques non-linéaires avec applications à l'élasticité.* Université Pierre et Marie Curie, Paris VI, Analyse Numérique, 1981.

[71] NEČAS, J.; STARÁ, J.; ŠVARC, R.: *Classical solution to a second order non linear elliptic system in R_3.* Ann. S.N.S. Pisa 5 (1978), 605–631.

[72] GIAQUINTA, M.; NEČAS, J.; JOHN, O.; STARÁ, J.: *On the regularity up to the boundary for second order nonlinear elliptic system.* Pacific J. Math. Vol. 98, N. 1, 1981.

[73] CAMPANATO, S.: *Equazioni ellitiche del secondo ordine e spazi* $\mathscr{L}^{2,\lambda}$. Ann. Mat. Pura e Appl. 69 (1965), 321–380.

[74] KRUŽKOV, S. I.: *Nelineĭnye uravnenija s častnymi proizvodnymi.* Matemat. sbornik t. 65 (107) n. 4, 1964, 522–570.

[75] GUO ZHONG-HENG: *The unified theory of variational principles in non-linear elasticity*. Arch. Mech. 32, 4 (1980), 577–596.

[76] CIARLET, P.: *Lecture notes.* 1981, Pierre et Marie Curie Université, Paris VI, Analyse Numérique.

[77] TRUESDELL, C.: *A first course in rational continuum mechanics.* J. Hopkins University, 1972.

Index